Artillery of the World

Christopher F. Foss

Charles Scribner's Sons
NEW YORK

To my wife Elaine

U.S. edition published by
Charles Scribner's Sons 1981

1 3 5 7 9 11 13 15 17 19 I/C 20 18 16 14 12 10 8 6 4 2

Printed in Great Britain

Library of Congress Catalog Card Number 80-53549
ISBN: 0-684-16722-0

Contents

Introduction

This, the third edition of *Artillery of the World* to be published since 1974, has been completely revised and contains much new information and many new photographs of ground and anti-aircraft artillery, multiple rocket systems and their associated fire control systems.

Since the second edition was published no less than three 155mm towed gun-howitzers have entered service, the American M198, the Swedish Bofors FH-77 and the International (Britain, West Germany and Italy) FH-70. Both Belgium and Israel have recently developed new 155mm gun-howitzers, although these are aimed primarily at the export market rather than the developing country concerned. In mid-1979 France showed the prototype of its new 155mm towed gun-howitzer which is expected to enter service in the early 1980s while South Africa showed its 155mm G-5 later in the year. Many new types of ammunition are now entering service with the United States Army such as the 155mm Rocket Assisted Projectile, 155mm Cannon Launched Guided Projectile, 155mm Remote Anti-Armour Mine System (carries nine anti-tank mines), 155mm Area Denial Artillery Munition (carries 36 anti-personnel mines), as well as other projectiles that carry anti-personnel or dual purpose grenades. These will enhance the effectiveness of artillery units, especially against armoured vehicles.

The multiple rocket system section has been rewritten and now includes details of the American Multiple Launch Rocket System (previously known as the General Support Rocket System) which should become the standard system of NATO in the 1980s.

There are many new entries in the fire control section, although by no means is every system covered and some of the older equipment has now been phased out of *Artillery of the World* to make way for the new entries.

Details of self-propelled guns and howitzers, tank destroyers and self-propelled anti-aircraft guns are given in the companion volume *Armoured Fighting Vehicles of the World* while details of many of the trucks and full track prime movers used to tow artillery are given in *Military Vehicles of the World*.

Additional or fresh material for future editions of *Artillery of the World* should be forwarded to *Artillery of the World*, Ian Allan Limited, Terminal House, Shepperton, Surrey TW17 8AS, England. New photographs of towed guns, howitzers, anti-aircraft guns and multiple rocket systems would be particularly welcome.

The author would like to take this opportunity of thanking the many governments, companies and individuals all over the world who have assisted in the preparation of this book. Special thanks are due to Mr T. Cullen, Mr K. Ebata, Mr T. Gander, Mr K. Nogi, Mr X. Taibo, Mr G. Tillotson and Dr A. Volz for their valuable assistance.

January 1980 **Christopher F. Foss**

Acknowledgements

The author is grateful for the help he has received from the following:

AEG-Telefunken, West Germany
Argentine MoD
Austrian MoD
B&W Elektronic, Denmark
Bofors, Sweden
Canadian Armed Forces
CILAS, France
CISMA, France
CNIM, France
Contraves Co, Italy and Switzerland
Egyptian MoD
Electronique Marcel Dassault, France
ELTA Electronics, Israel
EMI Limited, UK
Esperanza Y, Cia, SA, Spain
Ferranti Ltd, UK
Finnish MoD
French Army
General Electric Company, USA

German MoD
Gesellschaft für Ungelenkte Flugkörpersysteme
MBB, FRG
GIAT, France
Hollandse Signaalapparaten
Hughes Co, USA
Indian MoD
Israel Aircraft Industries, Israel
Israeli Defence Forces (General Staff)
Kongsberg Vapenfabrikk, Norway
L. M. Ericsson Company, Sweden
Marconi Space and Defence Systems Co Ltd, UK
Martin Marietta Aerospace Co, USA
MoD (Army), UK
Nera Bergen, Norway
Nissan Motor Co Ltd, Japan
N. V. Optische Industrie, Netherlands
Oerlikon-Bührle Co, Switzerland
OTO-Melara, Italy
Philips Teleindustri, Sweden
Plessey Co, UK
RAFAEL Armament Development Authority, Israel
Rheinmetall Co, FRG
Saab-Scania (Aerospace Division), Sweden
SAI, Italy
Saunders Associates, USA
SEP, France
Simrad as, Norway
Soltam Co, Israel
SRC International, Belgium
Swedish Army Material Department
Swiss MoD
Thomspon-Brandt (Armament Department), France
US Army
Vought, USA

Abbreviations

AA	Anti-Aircraft
AE	Atomic Explosive
AMETS	Artillery Meteorological System
APC	Armoured Personal Carrier Armour-Piercing Caped
APDS(T)	Armour-Piercing Discarding Sabot (Tracer)
APHE	Armour-Piercing High Explosive
API(T)	Armour-Piercing Incendiary (Tracer)
ATG	Anti-Tank Gun
aux prop	auxiliary-propelled
AWDATS	Artillery Weapon Data Transmission System
BATES	Battlefield Artillery Target Engagement System
CLGP	Cannon-Launched Guided Projectile
ECM	Electronic Countermeasures
ECCM	Electronic Counter-Countermeasures
ERFB	Extended Range Full Bore
FACE	Field Artillery Computer Equipment
FAAR	Forward Area Alerting Radar
FCS	Fire Control System
FDC	Fire Direction Centre
FG	Field Gun
FH	Field Howitzer
GH	Gun-Howitzer
GLLD	Ground Laser Locator Designator
g	gramme
h	hour
HE	High Explosive
HEAT	High Explosive Anti-Tank
HEI	High Explosive Incendiary
HEPT	High Explosive Plastic Tracer
HES	High Explosive Spotting
HESH	High Explosive Squash Head
HMG	Heavy Machine Gun
how	howitzer
HVAP	High Velocity Armour-Piercing
IFF	Interrogation (Identification) Friend or Foe
kg	kilogramme
km	kilometre
LAAG	Light Anti-Aircraft Gun
LARS	Light Artillery Rocket System
LFH	Light Field Howitzer
LTD	Laser Target Designator
LWB	Long Wheelbase
m/v	muzzle velocity
MG	Machine Gun Mountain Gun
MILIMETS	Military Meteorological System for Artillery
MORCOS	Mortar Data Computing System
MRS	Multiple Rocket System
MTBF	Mean Time Between Failures
MULE	Modular Universal Laser Equipment
PADS	Position and Azimuth Determining System
PH	Pack Howitzer
PPI	Position Plan Indicator
RAP	Rocket-Assisted Projectile
rd	round(s)
R/F	Rangefinder
RL	Rocket Launcher
ROF	Rate of Fire
rpm	rounds/min
RR	Recoilless Rifle
SP	Self-Propelled
TP(T)	Target Practice (Tracer)
w/o	without
WP	White Phosphorus

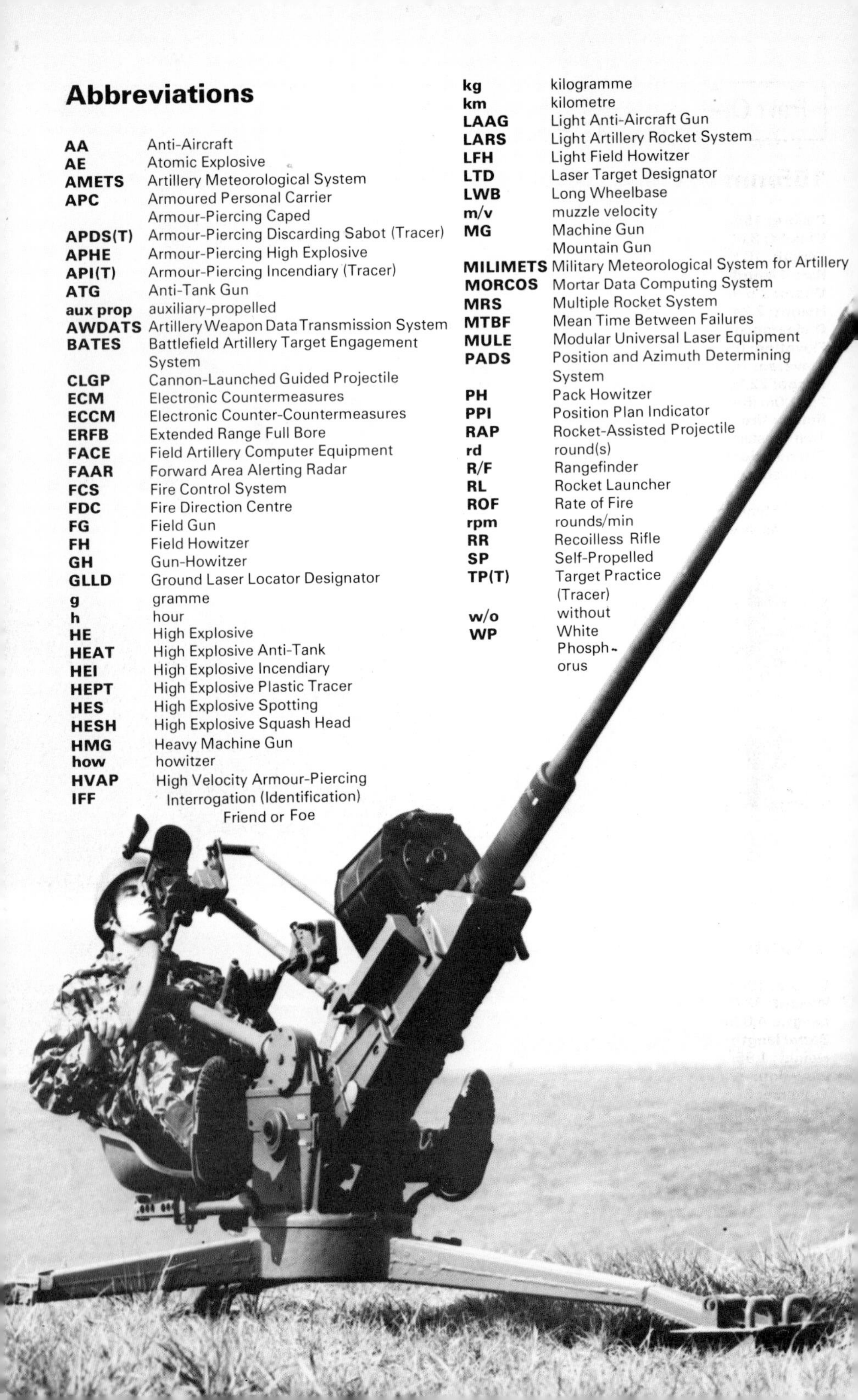

Part One: Guns and Mortars

155mm CITEFA L33 X1415 Model 77 Howitzer Argentina

Calibre: 155mm
Weight: 8,000kg (travelling)
Length: 10.5m
Barrel length: 5.115m
Width: 2.67m
Height: 2.2m
G/clearance: 0.3m
Elevation: 0° to +67°
Traverse: 70° (total)
Range: 22,000m (normal)
25,300m (RAP)
Rate of fire: 4rpm (max)
1rpm (sustained)
Ammunition: HE, projectile weight 43kg, m/v 765m/sec; illuminating; smoke

This 155mm howitzer has been designed by the Scientific and Technical Research Institute of the Armed Forces (CITEFA) to meet the requirements of the Argentinian Army. It is essentially the top carriage (barrel, cradle, recoil system and equilibrators) of the French AMX 155mm self-propelled howitzer Mk F3 mounted on a new split trail bottom carriage. The Mk F3 has been in service with the Argentinian Army for some years. When in the firing position the L33 is supported on the firing base under the carriage and the trails.

Employment
Argentina.

155mm howitzer L33 X1415 CITEFA Model 77 in firing position.

105mm Model 1968 Recoilless Gun Argentina

Calibre: 105mm
Weight: 397kg (complete)
Length: 4.02m
Barrel length: 3m
Height: 1.95m
Elevation: −7° to +40.5°
Traverse: 360°
Range: 700m (max effective)
Rate of fire: 12rpm
Ammunition: HEAT, weight 11.1kg, m/v 405m/sec
HE, weight 15.6kg, m/v 400m/sec
Armour penetration: 200mm (HEAT round)

105mm recoilless gun Model 1968 designed and built in Argentina, mounted on a 4×4 Unimog truck.

Spotting rifle: 7.62mm DGMF FAP; weight 6.4kg, max range 1,200m
Telescope: ×4 magnification, 12° field of view

This has been designed and manufactured in Argentina by the Direccion General de Fabricaciones Militares and is now in service in Argentina. In addition some have been exported to Peru. Its full designation is Canon Sin Retrocesco Calibre 105mm Modelo 1968. It is normally carried in the rear of a 4×4 Unimog truck and loading ramps are provided so that it can be quickly removed from the vehicle when required.

Argentina is now building both 81mm and 120mm mortars and 75mm recoilless rifles both for the home and export markets.

155mm GC45 Gun-Howitzer

Belgium

Calibre: 155mm
Weight: 8,222kg (travelling and firing)
Length: 13.614m (travelling, barrel forwards)
9.244m (travelling, barrel over trails)
Barrel length: 6.975m
Width: 2.692m (travelling)
Height: 3.28m (travelling)
G/clearance: 0.355m
Elevation: −5° to +69°
Traverse: 80° (total)
Range: 30,000m
Rate of fire: 2rpm (sustained)
4rpm (for 15min)
Ammunition: HE, weight 45.4kg, m/v 897m/sec; illuminating; smoke
Crew: 8
Towing vehicle: 5ton (6×6) truck

The GC45 gun-howitzer has been developed as a private venture by SRC International of Belgium, with the first prototype being completed in 1977.

The 45 calibre long barrel has been designed to fire all standard as well as new 155mm NATO ammunition and charges, with the same velocity and range as would be achieved with a 39 calibre barrel. The main advantage of the GC45 over the American M198 and the German/Italian/United Kingdom FH-70 is that the NATO requirement for a range of 30,000m is achieved with a completely ballistic solution, rather than a RAP. The range of 30,000m is achieved using the SRC developed ERFB Mk 10 mod 2 projectile with an M11 charge.

To enable a high rate of fire to be achieved a pneumatically operated telescoping rammer is provided. When in the firing position the GC45 is supported by a base plate under the forward part of the carriage and the two trails. When travelling the barrel can be traversed through 180° and locked in position over the trails, or locked in the normal firing position. The walking beam suspension enables the GC45 to be towed on roads at speeds of up to 90km/h.

Employment
Thailand (Marines).

SRC International GC45 155mm gun-howitzer in travelling configuration with ordnance forwards.

155mm M114/39 Modified Howitzer

Belgium

Calibre: 155mm
Weight: 7,024kg (travelling)
Length: 10.363m (travelling)
Width: 2.438m (travelling)
G/clearance: 0.229m
Elevation: −2° to +63°
Traverse: 25° (left) and 24° (right)
Range: 26,000m
Rate of fire: 4rpm
Ammunition: HE, projectile weight 45.5kg; illuminating; smoke
Crew: 11
Towing vehicle: 5ton (6×6) truck

The M114/39 modified howitzer has been developed as a private venture by SRC International for the export market. Development is now complete and SRC can either supply conversion kits to enable the user to convert existing M114s in his own workshops or convert the M114 for the customer in his own country or at SRC facilities.

The basic modification is to replace the existing M114 barrel with a new 39 calibre barrel with a larger chamber, faster twist rifling and a muzzle brake. The original equilibrators have been replaced and modifications have also been made to the recoil system, elevating and traversing mechanisms, carriage and trails. A swing aside loading tray and a pneumatic ramming system have been fitted, similar to those installed on the 155mm GC45 gun howitzer.

The M114/39 modified howitzer can fire a standard M107 HE projectile to a maximum range of 22,100m compared to the 14,600m range of the basic M114, or an SRC ERFB projectile weighing 45.4kg to a maximum range of 26,000m. SRC also offer a 45 calibre barrel which can fire the ERFB projectile to a range of 30,000m.

Employment
Ready for production.

SRC International M114/39 modified howitzer in towed configuration showing the air bottle that operates the rammer on the right trail.

81mm PRB Model NR475A1 Mortar — Belgium

Calibre: 81mm
Weight: 43kg (firing)
Weight of barrel: 15.3kg
Weight of bipod: 12.5kg
Weight of baseplate: 14.6kg
Weight of sight: 0.6kg
Barrel length: 1.35m
Elevation: +40° to +85°
Traverse: 8°
Range: 300m (min)
5,500m (max)
Rate of fire: 15-20rpm

The Model NR475A1 81mm mortar has been developed by the well known Belgian ammunition manufacturer PRB. The mortar will fire both American and PRB mortar bombs, the latter includes the NR414 HE (weight 3.25kg), NR436 long range HE, NR163 smoke (weight 5.8kg) and NR164 illuminating (weight 4.6kg).

The smooth bore barrel has radial fins on its lower half to assist in cooling, and the circular baseplate has four welded ribs and five spikes on the under side. The bipod legs are connected by a chain with the elevating mechanism being connected to the left leg.

Employment
Belgium.

105mm C1 Howitzer — Canada

The 105mm Howitzer C1 is essentially the American 105mm M101 howitzer built under licence in Canada by Sorel Industries Limited of Sorel, Quebec, in the mid-1950s. It remains in service with reserve units of the Canadian Armed Forces although it is possible that it may be reintroduced into regular units to replace their OTO-Melara 105mm pack howitzers.

40mm L40/60 Boffin AA Gun

Canada

The 40mm L40/60 Boffin anti-aircraft gun was originally used by the Royal Canadian Navy during World War 2 and consists of an Oerlikon twin 20mm Mark V(C) mount fitted with a 40mm Bofors gun. In the early 1970s they were taken out of store and deployed in the static role to defend two Canadian airfields in West Germany. The gun and mount weighs 1,770kg and the weapon has an elevation of +66°, a depression of −3°, a total traverse of 360°, and is manned by a three-man detachment.

China, People's Republic of

14.5mm Type 56 AA Gun
This is the Soviet ZPU-4 built in China.

14.5mm Type 58 AA Gun
This is the Soviet ZPU-2 built in China.

37mm Type 55 AA Gun
This is the Soviet M1939 built in China; a twin version is also produced under the designation of the Type 63 and a self-propelled model of the latter, based on the chassis of the T-34/85 tank, has been used by Vietnam under the designation of the Type 63 self-propelled anti-aircraft gun.

57mm Type 59 AA Gun
This is the Soviet S-60 built in China.

85mm Type 56 AA Gun
This is the Soviet KS-18 built in China.

100mm Type 59 AA Gun
This is the Soviet KS-19 built in China.

57mm Type 55 Anti-Tank Gun
This is the Soviet 57mm Anti-Tank Gun M1943 built in China.

57mm Type 36 Recoilless Rifle
This is the American 57mm M18 Recoilless Rifle modified by the Chinese and built in China. It has also been exported to a number of countries including Tanzania and Vietnam.

76mm Type 54 Anti-Tank Gun
This is the Soviet M1942 (ZIS-3) 76mm gun built in China. It has also been exported to a number of countries including Kampuchea, Tanzania and Vietnam. Full details are given in the Soviet section.

75mm Type 52 Recoilless Rifle
This is a modified United States M20 Recoilless Rifle built in China. Some have been supplied to Vietnam.

75mm Type 56 Recoilless Rifle
This is a modified Type 52 weapon.

82mm Type 65 Recoilless Rifle
This is a lighter version of the Soviet B-10 and is replacing the 75mm Type 56 recoilless rifle.

Chinese Type 55 light AA gun used by the Tanzanian Army.

82mm Type 53 Mortar
This is a modified M1937 Soviet mortar. It has also been exported to Kampuchea, Pakistan, Tanzania, Vietnam and Uganda. For full details refer to USSR section.

85mm Type 56 Field Gun
This is the Soviet D-44 built in China.

100mm Type 59 Field Gun
Unconfirmed copy of Soviet M1944 (BS-3).

120mm Type 53 Mortar
This is the Soviet 120mm M1943 mortar built in China. It has also been supplied to Pakistan, Tanzania and Vietnam. It has also been called the Type 55.

122mm Type 54 Howitzer
This is the Soviet 122mm M1938 built in China.

122mm Type 60 Field Gun
This is the Soviet 122mm D-74 built in China.

160mm Type 56 Mortar
This is the Soviet M1953 160mm mortar built in China.

130mm Type 59 Field Gun
This is a copy of the Soviet M46.

130mm Type 59-1 Field Gun
This Chinese-designed 130mm gun is mounted on the carriage of the 122mm Type 60 gun and fires a high explosive projectile weighing 33.33kg to a maximum range of 22,000m.

152mm Type 54 Howitzer
This is a copy of the Soviet M1943 weapon.

152mm Type 66 Gun-Howitzer
This is a copy of the Soviet D-20 weapon and is mounted on the carriage of the 122mm Type 60 gun.

152mm M18/46 Howitzer

Czechoslovakia

Calibre: 152mm
Weight: 5,512kg (firing)
Length: 8.284m (travelling)
Barrel length: 4.875m (inc muzzle brake)
Width: 2.255m
Height: 1.707m
Elevation: 0° to +45°
Traverse: 60° (total)
Range: 12,400m (HE round max)
Rate of fire: 4rpm
Ammunition: HE, projectile weight 39.9kg, m/v 508m/sec
Semi-AP, projectile weight 51.1kg, m/v 432m/sec
Armour penetration: 82mm at 500m (Semi-AP round)
Crew: 7
Towing vehicle: KrAZ-214 (6×6) truck
AT-S medium tracked artillery tractor

At the end of World War 2 the Czechoslovakians had a large number of the German 15cm sFH18 medium howitzers left behind by the Germans. The Czechs rebored the weapons to fire the Soviet 152mm howitzer round and also added a straight topped shield and a large double baffle muzzle brake.

It has a hand-operated horizontal sliding wedge type breechlock, hydraulic recoil buffer and a hydropneumatic recuperator. The carriage has split trails and solid tyred wheels. The ammunition is of the case-type, variable charge, separate-loading type.

Employment
Czechoslovakia (reserve).

Czech 152mm howitzer M18/46 in the firing position; note the shield and the wheels.

105mm M18/49 Howitzer

Czechoslovakia

Calibre: 105mm
Weight: 1,750kg (firing)
Length: 5.98m (travelling)
Barrel length: 3.308m (inc muzzle brake)
Width: 2.1m (travelling)
Height: 1.8m (travelling)
Track: 1.56m
G/clearance: 0.4m
Elevation: −5° to +42°
Traverse: 60° (total)
Range: 12,320m (HE round max)
Rate of fire: 6-8rpm
Ammunition: HE, projectile weight 14.8kg, m/v 540m/sec
Crew: 8
Towing vehicle: Praga V3S (6×6) truck
Tatra-111R (6×6) truck

This weapon is the German World War 2 M18 light field howitzer mounted on the modified carriage of the German PAK40, and known in the German Army as the leFH 18/40.

The Czechs fitted the weapon with single rubber tyres in place of the heavy spoked wheels that had solid tyres. The weapon has a double baffle muzzle brake, a hydraulic recoil buffer and a hydro-pneumatic recuperator, the breechblock is of the hand-operated horizontal sliding wedge type. The ammunition is of the case-type, variable-charge, separate-loading type.

Employment
Czechoslovakia (reserve)

Czech 105mm howitzer M18/49 in the firing position.

100mm M53 Field Gun

Czechoslovakia

Calibre: 100mm
Weight: 4,280kg (firing)
Length: 9.1m
Barrel length: 6.735m (inc muzzle brake)
Width: 2.36m
Height: 2.606m
Track: 1.98m
Elevation: −6° to +42°
Traverse: 60° (total)
Range: 21,000m (HE round max)
Rate of fire: 8rpm
Ammunition: APHE, projectile weight 15.9kg, m/v 955m/sec
HE, projectile weight 15.7kg, m/v 930m/sec
HEAT, projectile weight 12.2kg, m/v 995m/sec
Armour penetration: 170mm at 1,000m (APHE round)
380mm at any range (HEAT round)
Crew: 6
Towing vehicle: Tatra-138 (6×6) truck

The M53 is the Czechoslovakian equivalent of the Soviet 100mm field gun, and fires the same ammunition as the latter and has similar performance.

It has a semi-automatic horizontal sliding wedge type breechblock, a straight top to its shield, the sides of which slope to the rear, a long barrel with a double baffle muzzle brake, split trails and single tyres. The gun can be fitted with the APN-3-5 infra-red night sight.

Employment
Czechoslovakia

85mm M52 Field Gun

Czechoslovakia

Calibre: 85mm
Weight: 2,095kg (firing)
Length: 7.52m (travelling)
Barrel length: 5.07m (inc muzzle brake)
Width: 1.98m (travelling)
Height: 1.515m (travelling)
Track: 1.63m
G/clearance: 0.35m
Elevation: −6° to +38°
Traverse: 60° (total)
Range: 16,160m (HE round max)
Rate of fire: 20rpm
Ammunition: HE projectile weight 9.5kg, m/v 805m/sec
APHE projectile weight 9.3kg, m/v 820m/sec
HVAP projectile weight 5.0kg, m/v 1,070m/sec
Armour penetration: 123mm at 1,000m (APHE round)
115mm at 1,000m (HVAP round)
Crew: 7
Towing vehicle: Praga V3S (6×6) Truck
Tatra-111R (6×6) Truck

	Soviet 100mm M-1944 ATG	Czech 100mm M53 FG	Czech 85mm M52 FG
Breechblock	vertical sliding wedge	horizontal sliding wedge	vertical sliding wedge
Shield top	straight	straight	wavy
Wheels	dual	single	single
Trails	box section	box section	box section
Muzzle brake	double-baffle	double-baffle	double-baffle

This weapon is the Czech equivalent of the Soviet D-44 gun and fires the same type of fixed ammunition as the various Russian 85mm guns and has a similar performance. It can be mistaken for the Soviet or Czech 100mm guns. The M52 has stowage boxes on the trails. A modified version of this weapon is called the M52/55 and weighs 2,111kg in the firing position.

Employment
Austria (250), Czechoslovakia and East Germany.

85mm M52 field gun of the Austrian Army.

82mm M59A Recoilless Gun

Czechoslovakia

Calibre: 82mm
Weight: 386kg (firing and travelling)
Length: 4.28m (travelling)
Width: 1.675m (travelling)
Height: 1m (travelling)
Track: 1.52m
G/clearance: 0.325m
Height of Axis of Bore at 0° Elevation: 0.805m
Elevation: −13° to +25°
Traverse: 360° (at 0° elevation)
60° (at 25° elevation)
Rate of fire: 6rpm
Range: 7,560m (HE round max)
point blank 2m high target 755m (HEAT round)
Ammunition: HE projectile weight 6kg, m/v 565m/sec
HEAT projectile weight 6kg, m/v 745m/sec
Armour penetration: 250mm (HEAT round)
Spotting rifle: 12.7mm
Crew: 5
Towing vehicle: Tatra-805 (4×4) truck
UAZ-69 (4×4) truck
OT-810 half track

This weapon has the same calibre as the lighter T-21 but a much greater range, and was built by the Skoda works in Pilsen. The weapon is employed in the Czechoslovakian Army within the motorised rifle regiment as an anti-tank gun for infantry supporting fire using both direct and indirect laying.

It is normally towed by its breech end behind a truck or APC, it is also often mounted on a UAZ-69 truck, carried on the rear decking of an OT-62 APC or mounted on the top of an OT-810 halftrack. The muzzle end is provided with a bar to assist in manhandling the weapon.

The M59A is the only weapon in the Warsaw Pact Forces that uses a 12.7mm spotting rifle which is mounted over the barrel. The latest version is the M59A which can be distinguished from the earlier M59 by the ribbed configuration of the outside of the chamber. This version has a much greater range than the M59 and is fitted with the PBO-4K sight.

Employment
Czechoslovakia, Egypt, Poland.

82mm recoilless gun M59A in the firing position.

30mm M53 Twin AA Gun

Czechoslovakia

Calibre: 30mm
Weight: 2,000kg (in firing position)
2,100kg (in travelling position)
Length: 7.587m (travelling)
Barrel length: 2.429m (inc muzzle brake)
Width: 1.758m (travelling)
Height: 1.575m (travelling)
Elevation: −10° to +85°
Traverse: 360°
Range: 9,700m (max horizontal)
6,300m (max vertical)
3,000m (effective vertical)
Rate of fire: (per barrel) 450-500rpm (cyclic)
100rpm (practical)
Feed: 10-round clips fed horizontally
Ammunition: HEI projectile weight 0.45kg, m/v

1,000m/sec
API projectile weight 0.45kg, m/v 1,000m/sec
Complete round's weight 0.91kg
Armour penetration: 55mm at 500m
Crew: 4
Towing vehicle: Praga-V3S (6×6) truck

This weapon was first seen in 1958 and is used in a similar role as the Russian 23mm guns. The twin guns have hydraulic elevation and traverse and have quick change barrels. Mounted on a four wheeled carriage, a stowage box is on the right hand side of the mount. A self-propelled model of the M53 is also in service with the Czechoslovak Army. This model is designated the M53/59 and consists of an armoured Praga V3S (6×6) truck chassis with the twin 30mm weapons mounted on the rear. The weapons have a higher rate of fire than the towed version as ammunition is fed from 50-round magazines rather than the 10-round clips of the latter. The latest version of the M53/59 is designated the M53/70 and has an improved fire control system.

Employment
Czechoslovakia and Vietnam.

Twin 30mm AA gun M53; note the screw type firing jacks under the carriage.

57mm AA Gun — Czechoslovakia

Calibre: 57mm
Weight: 5,150kg (firing)
5,150kg (travelling)
Length: 5.5m (travelling)
Barrel length: 4m
Height: 2m (travelling)
Elevation: –5° to +87°
Traverse: 360°
Range: 12,000m (max horizontal)
8,800m (max vertical)
6,000m (effective vertical)
Rate of fire: 80-90rpm
Feed: 3-round clip
Ammunition: APHE projectile weight 3.1kg, m/v 1,005m/sec
HE projectile weight 2.58kg, m/v 1,000m/sec
Armour penetration: 106mm at 500m (APHE round)
Crew: 7
Towing vehicle: Praga V3S (6×6) truck

This was built in Czechoslovakia in very small numbers and was replaced by the Russian 57mm S-60 weapon. It is heavier than the latter and has a similar rate of fire but a lower muzzle velocity.

57mm AA gun in the firing position. Note that the left of the shield folds down and that the firing jack between the wheels is in position.

Employment
Cuba, Guinea, Mali.

12.7mm M53 Quad AA HMG — Czechoslovakia

Calibre: 12.7mm
Weight: 628kg (firing)
2,830kg (travelling)
Length: 2.9m (travelling)
Barrel length: 0.967m (w/o muzzle brake)
Width: 1.57m (travelling)
Height: 1.78m (travelling)
Track: 1.5m

Elevation: −7° to +90°
Traverse: 360°
Range: 6,500m (max horizontal)
5,500m (max vertical)
1,000m (effective vertical)
Rate of fire: (per barrel) 550-600rpm (cyclic)
80rpm (practical)
Feed: Belt, 50 rounds in each drum
Ammunition: API, projectile weight 49.5g, m/v 840m/sec
Armour penetration: 20mm at 500m (API round)
Crew: 6
Towing vehicle: Tatra-805 (4×4) truck
GAZ-69 (4×4) truck

This was built in Czechoslovakia and uses four Russian-designed M-1938/46 DShK heavy machine guns. It is used in a similar role to the Soviet ZPU-2 and ZPU-4 weapons.

It is recognisable by its four large drum magazines, muzzle brakes on the barrels, two-wheeled mount with levelling jacks at the front and rear. It is no longer used in Czechoslovakia.

Employment
Cuba, Egypt and Vietnam.

Quad 12.7mm AA HMG 53 in the travelling position.

122mm Tampella M/60 Field Gun — Finland

Calibre: 122mm
Weight: 8,500kg
Elevation: −5° to +50°
Traverse: 90° (total)
360° (with additional equipment)
Range: 25,000m
Ammunition: HE projectile weight 25kg, m/v 950m/sec
Towing vehicle: Russian AT-S medium tracked

122mm Tampella field gun M/60 being towed by a Sisu KB-45 (4×4) truck.

artillery tractor
Finnish Sisu KB-46 (6×6)

The 122mm Tampella M/60 field gun is of Finnish design and construction and is also used in the coastal defence role. The weapon is mounted on a heavy duty split trail carriage which has a total of four road wheels which can be driven by an hydraulic motor to improve its cross-country mobility when being towed by a truck. The long barrel (53 calibres) is provided with a muzzle brake but the weapon has no shield. When in the firing position a jack is lowered under the carriage to provide a more stable firing platform. When in the travelling position the barrel is swung through 180° so that it rests over the closed trails.

More recent Tampella weapons are the 155mm M-68 and the M-71 neither of which are in service with the Finnish Army. The M-68 howitzer has an elevation of +52° and a depression of −5° and fires a projectile weighing 43.6kg to a maximum range of 21,000m. The barrel length is 33 calibres. The M-71 howitzer has a 39 calibre barrel and fires a projectile weighing 43.6kg to a maximum range of 23,500m. Details of these are given under Israel.

Employment
Finland.

105mm M/61-67 Light Field Howitzer — Finland

Calibre: 105mm
Weight: 1,800kg
Elevation: +6° to +45°
Traverse: 53° (total)
Range: 13,400m
Rate of fire: 7rpm
Ammunition: HE weight 14.9kg, m/v 600m/sec
Towing vehicle: Sisu KB-45 (4×4) (Finnish)
Sisu KB-46 (6×6) (Finnish)

The M/61-67 is essentially the older 105mm Tampella light field howitzer fitted with a new 105mm Tampella gun. The carriage is of the split trail type with each trail being provided with a lifting handle. A small shield provides protection from small arms fire and shell splinters. A similar hybrid was constructed at the same time by mounting the gun of the Tampella light field howitzer on the carriage of a 122m Soviet M-10 to make the M-37/10 light field howitzer.

105mm light field gun M/61-67.

Other Artillery Weapons used by Finland

The Finnish Army has its own designations for foreign weapons used by its Army:

M/02 — A modified 76mm Russian gun.
M/34 — A 152mm Russian gun.
M/36 — A 76mm Russian gun.
M/38 — A 152mm howitzer of Finnish construction with a range of 12,000m, total weight is 4,200kg.
M/40 — The German World War 2 150mm Model 40 field howitzer.
M/41 — The Russian 105mm gun modified.
M/54 — The Soviet 130mm M-46 field gun.
M/37 — The German World War 2 88mm Flak AAG 36/37.
M/40 — The German World War 2 20mm Flak AAG 38.

95mm M/58 Recoilless Rifle — Finland

Calibre: 95mm
Weight: 140kg
Barrel length: 3.2m
Range: 700m (effective)
Rate of fire: 6-8rpm
Ammunition: HEAT, projectile weight 10.2kg, m/v 615m/sec
HE, projectile weight 12.7kg, m/v 400m/sec
Armour penetration: 300mm
Crew: 3

This weapon was designed and built by Valmet Oy

and is mounted on two small wheels with a stabilising leg at the front and rear. To assist in moving the weapon there is a carrying handle either side, just in front of the breech, and another one either side of the barrel towards the front. The sights are on the left side of the barrel.

Employment
Finland.

95mm recoilless rifle M/58 in the firing position.

81mm M/38 Mortar

Finland

Calibre: 81mm
Weight: 60kg
Range: 3,000m
Rate of fire: 20rpm
Ammunition: 3.5kg
Crew: 3-4

This is reported to be based on the German World War 2 80mm Mortar GrW34, although it is much different to look at. The Finnish mortar has a circular baseplate, the bipod has a chain connecting the legs. On the left leg is a handle to alter the traverse of the mortar, the handle to alter the elevation is as the top of the tripod. The sights are on the left side of the barrel.

Employment
Finland.

81mm mortar M/38 and crew.

81mm M71 Mortar

Finland

Calibre: 81.40mm
Weight: 55kg (firing)
Weight of mortar bomb: 4.2kg
Range: 5,000m
Rate of fire: 20rpm

The M71 mortar has been developed by Tampella to meet the requirements of the Finnish Army. The mortar consists of a barrel with a smooth exterior surface, circular dished baseplate, bipod mounting with a distancing chain and screw jack elevating mechanism, cross levelling device with coarse and fine adjustments, a sight unit mounted on the yoke, and shock absorbers.

Employment
Finland.

120mm M73 Mortar

Finland

Calibre: 120.25mm
Weight: 236kg (firing)
Weight of mortar bomb: 12.8kg
Range: 8,000m
Rate of fire: 15rpm

The M73 mortar has been developed by Tampella to meet the requirements of the Finnish Army and is virtually a scaled up model of the 81mm mortar M71.

Employment
Finland.

160mm Tampella Mortar

Finland

This is used by the Finnish Army, has a weight in the firing position of 1,700kg and fires a 40kg bomb to a range of 9,500/13,800m. Rate of fire is 3-4rpm. It is also built in Israel in a much modified form.

For the complete range of Tampella Mortars the reader is referred to the Israel section under Soltam.

155mm Towed Gun

France

Calibre: 155mm
Weight: 9,500kg (travelling)
Length: 10m (firing)
8.5m (travelling)
Barrel length: 6.2m
Width: 2.93m (travelling)
Elevation: −5° to +66°
Traverse: 65° (total)
Range: 25,300m (standard ammunition)
30,500m (RAP)
Rate of fire: 3rpm (for first 15sec)
6rpm (for first 2min)
120 rounds/h
Ammunition: HE, projectile weight 43.2kg, m/v 810m/sec (range 25,300m)
Illuminating, projectile weight 43.75kg (range 21,500m)
Smoke, projectile weight 44.25kg, m/v 760m/sec (range 21,300m)
Brandt RAP, weight 42.5kg (range 30,500m)
Crew: 8
Towing vehicle: Berliet GBD (6×6) truck

This 155mm towed gun, called Le Cannon de 155mm Tracte, has been developed as the replacement for the M-1950 155mm field howitzer which has been in service with the French Army for some 25 years. It was shown for the first time in

The new 155mm French towed gun which was shown for the first time at the 1979 Satory Exhibition of Military Equipment.

June 1979 and is expected to enter production in the early 1980s.

The barrel has a double-baffle muzzle brake and when in the travelling position this is swung through 180deg and locked in position over the closed trails. The barrel is a development of that used in the 155mm GCT self-propelled gun which is based on the chassis of the AMX-30 MBT. Mounted on the forward part of the carriage is an engine which drives three hydraulic motors, one to provide power for elevation, traverse, raising the suspension, trail wheel jack and the projectile loading mechanism, while the other two motors power the two main road wheels. Maximum speed when being used in the self-propelled mode is 4km/h.

Employment
Development. Not yet in production or service.

155mm M-1950 Field Howitzer — France

Calibre: 155mm
Weight: 9,000kg (travelling)
Length: 7.15m (firing)
7.8m (travelling)
Width: 6.8m (firing)
2.75m (travelling)
Height: 1.65m (firing)
2.5m (travelling)
Elevation: $-4°$ to $+69°$
Traverse: 40° left and right
Range: 17,600m
Rate of fire: 4-5rpm
Ammunition: HE, projectile weight 43kg, m/v 650m/sec
Smoke, projectile weight 44.35kg
Illuminating, projectile weight 44kg
Towing vehicle: Berliet GBU 15 (6×6) truck (French Army)
Volvo TL31 L3153 (6×6) truck (Swedish Army)

The M-1950 155mm field howitzer is the standard 155mm howitzer of the French Army and has also been manufactured under licence in Sweden by Bofors, the Swedish designation being the 155mm Model F Field Howitzer. The weapon is mounted on a four-wheeled split carriage which is not provided with a shield. The barrel has a multi-baffle muzzle brake. When in the firing position a jack is lowered under the carriage to provide a more stable firing platform.

A self-propelled model of the M-1950 is in service with the French Army under the designation of the 155mm self-propelled howitzer Mk F3. This consists of a modified AMX-13 chassis with the howitzer mounted at the rear of the hull, the self-propelled model is also used by Argentina, Chile, Ecuador, Kuwait, United Arab Emirates and Venezuela. The self-propelled version, however, does have a long barrel as well as a double-baffle muzzle brake rather than the multi-baffle muzzle brake of the towed M-1950.

The Israeli Army also has a self-propelled version of the M-1950 towed howitzer, this being based on the chassis of a Sherman tank or a Priest self-propelled howitzer. The towed M-1950 was widely used by the Israeli Army but this has now been withdrawn from service.

Employment
France, Lebanon, Sweden and Tunisia.

155mm field howitzer M-1950 of the Israeli Army in the firing position.

20mm Cebere Twin AA Gun

France

The Cebere twin 20mm light anti-aircraft gun is basically the West German Rheinmetall anti-aircraft gun system with the original Rh202 cannon replaced by French M693(F2) cannon. It has been adopted for airfield defence by the French Air Force under the designation of the 76T2. Data on the Cebere is identical to the German weapon except that its travelling weight without ammunition is 2,100kg and its weight in the firing position with ammunition is 1,600kg.

Employment
In service with French Air Force.

Cebere twin 20mm AA gun in the firing position.

20mm Centaure Twin AA Gun

France

Calibre: 20mm
Weight: 994kg (firing and travelling, with ammunition)
914kg (w/o ammunition)
Length: 4.815m (travelling)
Barrel length: 2.065m
Height: 1.8m (travelling)
Track: 1.63m
Elevation: –5° to +80°
Traverse: 360°
Range: 1,500-2,000m
Rate of fire: (per barrel) 740rpm (cyclic)
200rpm (practical)
Crew: 3 (1 on gun)
Towing vehicle: Jeep or similar vehicle

Centaure twin 20mm AA gun in the firing position.

The Centaure is a joint development for the export market between the French Groupement Industries des Armements Terrestres and the Spanish Centro de Estudios Tecnicos de Materiales Especiales (CETME) and consists of a Spanish carriage and mount with its original 20mm guns replaced by French 20mm M693(F2) guns as used in the Cebere (2 × 20mm) and Tarasque (1 × 20mm) systems.

Each of the 20mm guns has a ready use supply of 100 rounds of ammunition with a flexible chute containing 15 rounds of AP ammunition. The gunner can select either single shots or full automatic and a wide range of ammunition can be fired including APDS (m/v 1,300m/sec), HEI and HEI-T (m/v 1,050m/sec) and TP and TP-T. Elevation and traverse are both manual with one handle being provided for elevation and one for traverse.

Employment
Ready for production.

20mm Tarasque AA Gun

France

Calibre: 20mm
Weight: 660kg (firing)
840kg (travelling)
Barrel length: 2.065m
Width: 1.9m
Track: 1.72m
Elevation: −8° to +83°
Traverse: 360°
Range: 1,500-2,000m
Rate of fire: 740rpm (cyclic)
200rpm (practical)
Crew: 3 (1 on gun)
Towing vehicle: Jeep or similar vehicle

The Tarasque 20mm anti-aircraft gun has been developed by the Groupement Industriel des Armaments Terrestres to meet the requirements of the French Army who have adopted the weapon under the designation of the 53T2. The Tarasque uses the same 20mm M693(F2) gun as the Centaure (2 × 20mm) and Cebere (2 × 20mm) anti-aircraft systems and this is also installed in a number of AFVs including the French AMX-10P IFV.

The 20mm gun is of the dual feed type with 100 rounds of HEI/HEI-T(m/v 1,050m/sec) and 40 rounds of APDS (m/v 1,300m/sec) being provided for ready use. Elevation and traverse is hydraulic with manual controls being provided for emergency use. The gunner can select either single shots or full automatic fire. The optical sight has a magnification of ×1 for use in the anti-aircraft role and ×5 for use in the ground role. The weapon takes only 15sec to bring into action from the travelling position.

Employment
French Army.

Tarasque 20mm AA gun in the firing position.

120mm Brandt Mortars

France

		MO-120-M65	*MO-120-RT-61*	*MO-120-AM50*	*MO-120-M60*
Weight:	inc carriage	144kg	580kg	402kg	—
	firing	104kg	580kg	242kg	92kg
	tube and breech	44kg	114kg	76kg	34kg
	mount	24kg	257kg	86kg	25kg
	baseplate	36kg	190kg	86kg	33kg
	carriage	40kg	—	154kg	—
Barrel length:	inc breech	1.64m	2.08m	1.746m	1.632m
Elevation:		+40° to +85°	+28° to +85°	+45° to +80°	+40° to +85°
Traverse:	(mils)	300	250	300	300
Rate of fire:	max (rpm)	12	10	12	15
	average (rpm)	8	6	8	8

The above mortars are all designed and manufactured by the French Company Hotchkiss Brandt.

M-120-M65

This is transported on a two-wheeled carriage, which has small pneumatic wheels and elastic suspension, the carriage is removed for firing. It is towed muzzle first by a light vehicle or two men. For transportation over rough ground it can be broken down into three man-pack loads. It is recognisable by its triangular baseplate and small wheeled carriage. The mortar can fire the following bombs:

PEPA/LP

Weight 13.42kg; max range with charge 6 9,000m; min range 1,200m. PEPA stands for Projectile Empenne à Propulsion Additionelle (Additional Propulsion).

M 44 HE

Weight 13kg; range 6,650m, min range 500m. Seven charges.

120mm Brandt mortar MO-120-M65.

Illuminating Mk 62

Weight 13.60kg; max range 5,000m; min range 900m.

Smoke

Weight 13kg; max range 6,650m; min range 500m. Also marking bombs and practice bombs, characteristics similar to the M44.

MO-120-RT-61

This is a rifled mortar and when in firing position its wheels are retained in position. The weapon consists of three parts: barrel with breech and towing ring, mount (cradle and undercarriage), and baseplate (triangular). It fires the following bombs:

PRPA (Rocket Assisted)

Total weight 18.7kg; max range 13,050m. This has additional propulsion.

RP-14

Weight 18.70kg; max range 8,135m; min range 1,200m. The weapon can also fire the M44, PEPA, PEPA/LP, Smoke and Illuminating bombs.

120mm Brandt rifled mortar MO-120-RT-61.

PRECLAIR (Projectile rayé éclairant)
This is an illuminating bomb and its shell weight is 15.45kg. Burning time is 60sec, and its candlepower for that burning time is of 1,050,000 candles or 40sec with 1,600,000 candles. Its diameter of illumination is of 1,000m. The min range is 300m and max range 8,000m.

120mm rifled anti-tank bomb
This has been designed for use with the 120mm rifled mortar MO-120-RT-61 and weighs 18.7kg and has a max range of 13,000m with rocket assistance.

MO-120-AM50
This weapon can be towed by its muzzle and can be fired with or without its wheels. It is recognisable by its two-wheeled carriage and triangular baseplate; when in the travelling position the barrel is in a horizontal position. Fires the PEPA/LP, M44, Smoke bomb Mk 62 and Illuminating Mk 62.

120mm Brandt mortar MO-120-AM50 in the firing position.

MO-120-LT
This is the latest Hotchkiss-Brandt mortar and is the replacement for the much heavier MO-120-AM 50. The MO-120-LT fires the same mortar bombs as the latter weapon but weighs only 203kg complete with its two-wheeled carriage. Weights of individual components of the mortar are: barrel 65kg, mount 25kg, baseplate 80kg, travelling carriage 48kg and barrel clamp and towing eye 7kg.

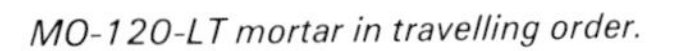

MO-120-LT mortar in travelling order.

MO-120-60
This weapon breaks down into three loads for easy transportation, ie tube and breech, bipod, and baseplate. There is also a special trailer to carry the mortar and 12 rounds of ammunition. It is recognisable by its star-shaped baseplate, bipod with chains connecting the two legs, twin short recoil cylinders under the barrel and the anti-cant device attached to the left leg. Fires the M44 bomb but only up to charge 4, PEPA/LP bomb but only to charge 4, Illuminating bomb Mk 62 but to charge 4 only, PEPA bomb (min 600m, max 6,550m range), also smoke, practice, target indicating, and coloured.

Employment
Used by many countries and also built under licence overseas.

These mortars are also mounted in various vehicles, for example the French AMX VCI has been fitted with a 120mm mortar.

120mm Brandt light mortar MO-120-60.

81mm Brandt Light Mortar

France

Calibre: 81mm
Weight: 39.40kg (short barrel)
41.50kg (long barrel)
Barrel weight: 12.40kg (short barrel)
14.50kg (long barrel)
Weight of baseplate: 14.80kg
Weight of mount: 12.20kg
Barrel length: 1.15m (short barrel)
1.45m (long barrel)
Elevation: +30° to +85°

This mortar is produced by Hotchkiss Brandt. There are two models — the short version (MO-18-61C) or the long version (MO-81-16-L). The weapon can be quickly broken down into three loads for transportation, ie barrel, baseplate and mount. The weapon is recognisable by its light, triangular shaped baseplate and bipod with chains connecting the legs. The 81mm mortar can also be mounted on vehicles, ie the AMX VTT APC.

The weapon can fire the following types of bombs:

HE Bomb Type M57D
Weight 3.30kg, max range 4,100m; min range 120m. Filling is TNT, this bomb being for the short version.

HE Bomb Type Mk 61
Weight 4.325kg; max range 5,000m; min range 75m; designed for the long-barrelled version although it can also be used for the shorter version. TNT filling.

Coloured HE Bombs M57D or Mk 61
Colours are red, green or yellow.

Smoke Bombs M57D or Mk 61
Data similar to HE bombs.

Practice Bombs Type M57D or Mk 61
Data similar to HE bombs. Filling is a mixture of sulphur and naphthalene with dummy fuzes.

Illuminating Bombs
Mk 62: weight 3.150kg; range 500-3,400m; 250,000 candlepower; min burning period 40sec.
Mk 68: weight 3.50kg; range 600-4,200m; 400,000 candlepower, min burning period 40sec.
Both have a filling based on magnesium and a lighting radius of 250m.

81mm light mortar long-barrelled version.

81mm light mortar short-barrelled version.

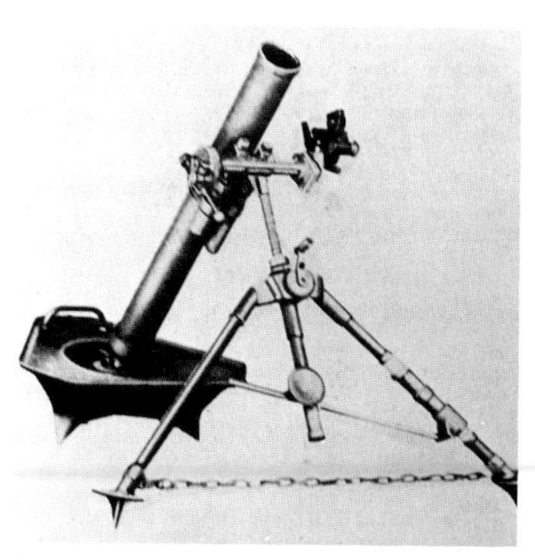

Employment
Short-barrelled version used by the French Army. Both long and short versions are in service outside France.

105mm M18 Light Field Howitzer

Germany

Data: Basic M18
Calibre: 104.9mm
Weight: 1,985kg (travelling)
Length: 5.994m (firing)
6.1m (travelling)
Barrel length: 2.706m
Width: 3.58m (firing)
1.977m (travelling)
Height: 1.88m (travelling)
Elevation: −5° to +42°
Traverse: 56° (total)
Range: 10,675m (max)
Rate of fire: 6-8rpm
Ammunition: HE, projectile weight 14.8kg, m/v 470m/sec
Crew: 6-10

Towing vehicle: Truck

The 105mm M18 (or to give the weapon its official German Army designation, the 10.5cm leichte Feldhaubitze 18) was developed in the late 1920s and entered service with the German Army in 1935. The M18 has a carriage of the split trail type, shield, light alloy road wheels with solid rubber tyres, hydraulic recoil buffer and a hydropneumatic recuperator. The breech mechanism is of the horizontal sliding wedge type and the barrel is not fitted with a muzzle brake.

The M18 was followed in 1940 by the M18(M) which is basically the earlier M18 with a modified recoil system and a double-baffle muzzle brake. This enabled the weapon to fire a more powerful charge and increased the range of the weapon from 10,675m to 12,325m.

The M18/40 was introduced in 1942 to meet a requirement for a weapon with the same range as the M18(M) but lighter to enable it to be manoeuvred more easily. The M18/40 is basically the carriage of the 75mm PAK40 anti-tank gun fitted with the barrel and recoil system of the M18(M). This has the same range as the M18(M) but weighs only 1,800kg.

Employment
Argentina (M18), Austria (M18/40), Chile (M18), Czechoslovakia (M18/49 qv), Portugal (M18), Sweden (M18 is called the m/39) and Yugoslavia (M18, M18(M) and M18/40).

105mm M18(M) with spades in the travelling position.

Germany World War 2

20mm Flak 38V AA Gun
This consists of four 20mm guns mounted on a quadruple mounting. They were on a two-wheeled trailer which was provided with levelling jacks; it was introduced into service in 1941. Basic data was as follows:
Weight: 2,212kg (travelling)
1,514kg (firing)
Length: 4.332m
Width: 2.336m
Height: 2.166m
Elevation: –10° to +100°
Traverse: 360°
Range: 4,800m (max horizontal)
3,700m (max vertical)
Rate of fire: 800rpm (practical)
Rounds carried: 80
Ammunition: HE, m/v 900m/sec; AP, m/v 830m/sec
There is also a single version of this weapon in service with Yugoslavia.

88mm Flak 18, 36 and 37 Multi-role Gun
This is the world-famous 88 gun which was developed in many versions. The 18, 36 and 37 were all mounted on carriages with four wheels and outriggers. Brief data as follows:
Weight: 5,600kg (firing)
7,800kg (travelling)
Length: 7.7m (travelling)
Height: 2.59m (travelling)
1.6m (firing)
Width: 1.6m (travelling)
Elevation: –3° to +85°
Traverse: 360°
Range: 14,860m (max horizontal)
10,600m (max vertical)
Rate of fire: 15-20rpm
Ammunition: Included HE, m/v 810m/sec; HVAP, m/v 935m/sec
The 88mm gun is still known to be used in Yugoslavia, mainly in the coast defence role.

15cm sFH18 Field Howitzer

This weapon has a hydropneumatic recoil mechanism and a horizontal sliding breechblock mechanism. When travelling a two-wheeled limber is attached to the rear and the barrel is moved out of battery. The sFH18 weighs 5,412kg when in the firing position and fires an HE projectile weighing 43.5kg to a max range of 13,325m. The weapon is still used by Finland (known as the m/40 and much modified) and Czechoslovakia. The Czechs have fitted it with a new barrel and call it the 152mm howitzer M18/46. For full details of this see the Czechoslovak section. Actual calibre of the sFH18 was 149mm.

20mm Mk 20 Rh202 Twin Cannon — Germany, Federal Republic of

Calibre: 20mm
Weight: 1,640kg (firing, inc ammunition)
2,160kg (travelling, inc ammunition)
Length: 5.035m (travelling)
Barrel length: 1.84m (excl muzzle brake)
Width: 2.36m (travelling)
Height: 2.075m (travelling)
Track: 2.08m
Elevation: −5° to +83°
Traverse: 360°
Range: 1,000-2,000m (tactical)
7,000m (max)
Rate of fire: 800-1,000rpm/barrel (cyclic)
Ammunition: HEI-T m/v 1,045m/sec
APDS-T m/v 1,150m/sec
API m/v 1,100m/sec
TP-T m/v 1,050m/sec
Armour penetration: AP round at 1,000m
NATO angle 0° 32mm armour plate
NATO angle 30° 24mm armour plate
NATO angle 60° 8mm armour plate
Ammunition supply: 270rd/barrel
Crew: 1 man on gun
2 men on replenishing ammunition
Towing vehicle: Mercedes-Benz Unimog (4×4) truck

This anti-aircraft system has been designed and manufactured by Rheinmetall for the German Army and Air Force. The weapon system consists of 2×20mm guns with their ammunition supply mounted on a cradle. This is mounted on the upper chassis which is in turn mounted on a lower chassis. This lower chassis has three outriggers, of these, two are adjustable for height and they support the weapon when firing. The lower chassis is mounted on a two-wheeled carriage for transportation.

The upper chassis also carries the gunner's seat, laying mechanism (including an air-cooled single-cylinder Wankel petrol engine which develops 8hp at 4,500rpm, hydrostatic drives for elevation and traverse, hydraulic dampers, oil tank etc, should this fail the guns can be laid manually), and the fire control system. The latter consists of a lead computer, target tracking device, the optical sight and the drive for the lead marks in the sight. It also has a 'taboo' feature so that friendly targets will not be fired upon.

The weapon has hydraulic elevation and traverse, traverse being 80°/sec and elevation 48°/sec, and can also be traversed and elevated manually. There is a selector switch for single shots or bursts, on both guns or each gun. Electrical or mechanical firing.

The Mk 20 Rh202 guns are also used in various other roles, including single 20mm anti-aircraft system (with mount being designed and built in Norway for the Norwegians), single 20mm shipboard mount, it is also mounted in a number of armoured fighting vehicles including the Marder MICV, Luchs (8×8) reconnaissance vehicle and the Fiat 6614 (4×4) armoured car.

Employment

Federal German Army (single mount) and Air Force (twin mount), Greek Army (twin mount), and Norway (single mount); it has also been built under licence in Italy by Whitehead Moto Fides at Leghorn, Italy.

Twin 20mm Mk 20 Rh202 system in the firing position.

Indian Artillery Developments

75mm Pack Gun/Howitzer
This weapon has been developed for use in mountain terrain and can probably be disassembled for transportation by pack animal. The carriage is of the conventional split trail type and the ordnance is fitted with a single-baffle muzzle brake. If required the wheels can be removed when in the firing position with the forward part of the carriage resting on a firing jack.

75mm pack gun-howitzer in the firing position with wheels removed.

105mm Light Gun
Early in 1978 the Jabalpur gun factory commenced the production of an Indian-designed 105mm light gun which, although weighing 3,500kg, is a lot heavier than the comparable British 105mm light gun which weighs 1,860kg. Maximum range is quoted as being 17,400m and rate of fire is 5rpm.

This will be produced only in limited numbers pending the production of the 105mm light field gun. This weapon, which is at present at the prototype stage, weighs about 2,000kg and also has a range of 17,400m. Like the British 105mm light gun, the Indian weapon has a circular platform to enable the weapon to be quickly traversed through a full 360°, and for travelling the ordnance is traversed through 180° so that it rests over the trails.

Other Weapons
India also manufactures 81mm and 120mm mortars as well as the American-designed 106mm M40 recoilless rifle.

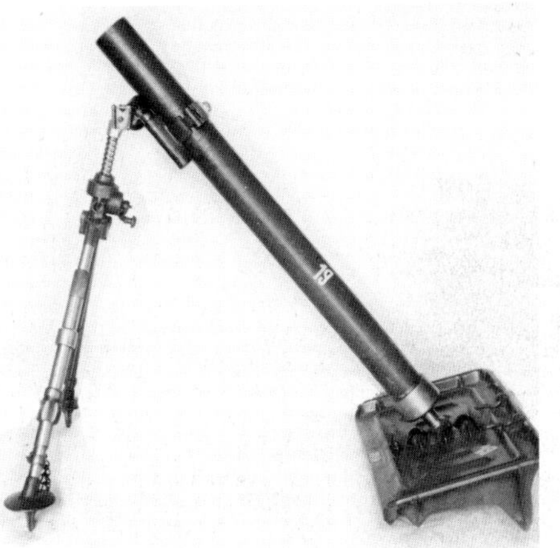

Indian-built 81mm mortar in the firing position.

Indian-built 120mm Brandt mortar in the firing position.

155mm FH-70 Field Howitzer

International

Calibre: 155mm
Weight: 9,300kg (firing)
9,300kg (travelling)
Length: 12.43m (firing)
9.8m (travelling)
Barrel length: 6.022m
Width: 2.2m (travelling)
Height: 2.195m (firing)
2.45m (travelling)
Track: 2.192m
G/clearance: 0.3m (travelling)
Elevation: $-5\frac{1}{2}°$ to $+70°$
Traverse: 55° (total)
Range: 24,000m (max, standard ammunition)
30,000m (max, RAP)
Rate of fire: 3rd in 13sec
6rpm (normal)
120rd/h
Ammunition: HE, BE smoke and illuminating projectiles, all of which weigh 43.5kg
Crew: 8
Towing vehicles: West Germany, MAN (6×6) truck
Italy, FIAT 6605 TM (6×6)
United Kingdom, Foden Medium Mobility Vehicle (6×6)

Development of the 155mm FH-70 (Field Howitzer 70) commenced in the late 1960s by West Germany and the United Kingdom with the latter being the project leader. The first of 19 prototypes were completed in 1969 and the following year Italy joined the project as a full partner. In 1975 a trials battery of six FH-70s was formed with two guns each being manned by men from each of the three countries. In 1976 FH-70 was accepted for service and first production weapons were completed in 1978. Prime contractors are Rheinmetall in West Germany, OTO-Melara in Italy and Vickers Shipbuilding Group in the UK. Under development is the self-propelled version of the FH-70, this is called

FH-70 being towed by a Foden (6×6) medium mobility vehicle.

The FH-70 in the firing position.

the SP-70 and is also being developed by West Germany, Italy and the UK with West Germany being the project leader. SP-70 will replace the American designed 155mm M109s used by all three countries but is not expected to enter service until the late 1980s.

The barrel has a double-baffle muzzle brake and a wedge breech mechanism. When travelling the barrel is traversed through 180° and locked in position over the closed trails. Mounted on the forward part of the carriage is an 1,800cc Volkswagen engine which enables the weapon to be deployed under its own power, max speed in the self-propelled mode being 16km/h. The APU also provides power for steering, raising and lowering both the main and trail wheels, a hand pump is provided for emergency use. The British Army designation for FH-70 is the L121 while West Germany calls it the FH155-1.

Employment
In service with the FGR, Italy and the UK.

155mm Soltam M-71 and M-68 Gun-Howitzer — Israel

Model:	*M-71*	*M-68*
Calibre:	155mm	155mm
Weight: (travelling)	9,200kg	9,500kg
Length: (travelling)	7.5m	7.2m
Barrel length:	6.045m	5.18m
Width: (travelling)	2.58m	2.58m
Height: (travelling)	2m	2m
Track:	2.2m	2.2m
G/clearance:	0.38m	0.38m
Elevation:	−5° to +52°	−5° to +52°
Max range:	23,500m	21,000m
Rate of fire: (short period)	4rpm	4rpm
(sustained)	2rpm	2rpm
Crew:	8	8
Towing vehicle:	5ton (6×6) truck	5ton (6×6) truck

Soltam 155mm M-71 gun-howitzer in travelling order.

The 155mm M-71 gun-howitzer has been developed from the earlier M-68 and uses the same carriage, recoil system and breech as the earlier weapon. Main difference is that the M-71 has a longer barrel (39 calibre) and has a compressed air rammer which enables the gun to be rapidly loaded at all angles of elevation. The carriage is of the split trail type and when travelling the ordnance is swung through 180° and locked in position over the closed trails. The ordnance has a single-baffle muzzle brake, fume extractor, hydropneumatic recoil system and a horizontal wedge breech mechanism. Both weapons fire an HE projectile weighing 43.7kg and can fire all standard NATO 155mm projectiles. Tampella designed projectiles can also be fired using a nine-charge system, in the case of the M-68, these enable a range of 23,500m to be achieved.

A self-propelled model of the M-68 is in service with the Israeli Army, this is called the L-33 and is based on a modified Sherman chassis. A self-propelled model of the M-71 has been developed to the prototype stage, this is based on a modified Centurion MBT chassis although the turret has been designed to be installed on other tank chassis such as the M48 or M60.

Employment
The M-68 is known to be in service with Singapore and Thailand.

Soltam 155mm M-68 gun-howitzer in travelling order.

20mm TCM-20 Light AA Gun — Israel

Calibre: 20mm
Weight: 1,350kg (travelling)
Length: 3.27m (travelling)
Width: 1.7m (travelling)
Height: 1.63m (travelling)
G/clearance: 0.31m
Elevation: –10° to +90°
Traverse: 360°
Range: 1,500m (effective AA)
Rate of fire: 650/700rpm/barrel (cyclic)
150rpm/barrel (practical)
Crew: 1 (on gun)
Towing vehicle: Jeep or other similar vehicle

The TCM-20 light anti-aircraft gun was developed by the MBT Division of Israel Aircraft Industries from 1969 to meet the requirements of the Israeli forces and is currently in service in two models. These are self-propelled with the mount being mounted on the rear of a modified American half-track and towed.

The TCM-20 is basically the American M55 anti-aircraft gun system, which was developed during World War 2, modernised with the original four 12.7mm M2 HB Browning machine guns being replaced by two Hispano Suiza HS804 20mm cannon with each being provided with a 60-round drum of ready use ammunition.

Elevation and traverse is electric with power being provided by two 12V batteries which are mounted at the rear of the turret, elevation and traverse speed is 60°/sec. The gunner aims the weapons with the aid of a standard M18 optical sight. When in the firing position the wheels are removed and the carriage is supported on three jacks, one at the front and two at the rear.

Employment
Israel.

20mm TCM-20 light AA gun in travelling order.

160mm Soltam M-66 Mortar

Israel

Calibre: 160mm
Weight: 1,450kg (travelling, w/o baseplate)
250kg (baseplate)
1,700kg (total firing)
1.45kg (sight Type A)
1.57kg (sight Type B)
Barrel length: 2.85m
Elevation: +43° to +70°
Traverse: 360°
Range: 9,300m (max)
Crew: 6-8

This smooth bore mortar consists of two main components, the carriage complete with the barrel and the baseplate. The weapon is towed without its baseplate by a standard light truck and can be quickly brought into action.

The carriage incorporates the barrel with breech piece, recoil-buffer, elevating mechanism and slow motion traversing gears. The mortar is fired by a lanyard attached to the firing lever. The folding elevating gear permits the barrel to lower itself automatically after the weapon has been fired. The counter-balance mechanism counteracts the weight of the barrel and facilitates the raising of the barrel after the bomb is loaded. When in the firing position the travelling wheels are turned inwards. The tangent position of the wheels permit the mortar to be traversed through 360° by using the wheels or the traversing mechanism which is situated on the left travelling wheel. The baseplate is of all-welded construction and is provided with four carrying handles.

The HE bomb weighs 40kg of which 5kg is explosive, max range is 9,300m and an average rate of fire is 5-8rpm. There are two types of smoke bomb, the $TiCl_4$ (Titanium Tetrachloride) and the Plastic Phosphorus, both of these have the same range as the HE bomb.

The $TiCl_4$ bomb produces instantaneous smoke whilst the Plastic Phosphorus bomb is a dual purpose bomb and can be used as a smoke bomb or as a night ranging bomb.

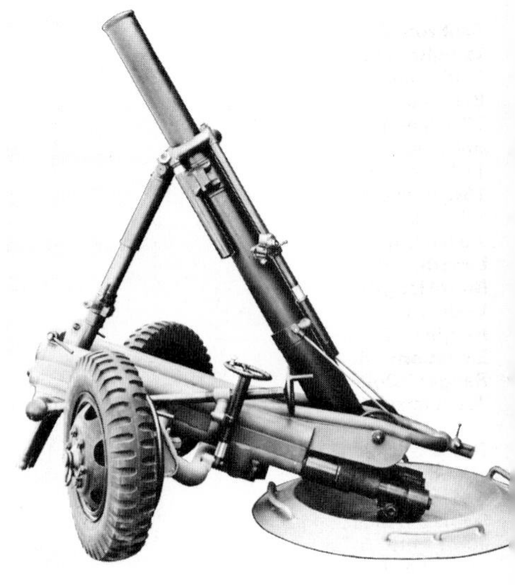

160mm Soltam M-66 mortar in travelling position.

The propellant system is made up of a primary charge and a combination of up to nine secondary charges. The secondary charges are circular in shape and are positioned above the tail unit.

Employment
In service with the Israeli Army. The Israeli Army has a number of these mounted on modified Sherman tank chassis.

160mm Soltam M-66 mortar in travelling order.

120mm Soltam M-65 Mortar

Calibre: 120mm
Weight: 346kg (total)
220kg (firing)
82kg (barrel and breech ring)
72kg (baseplate)
66kg (bipod)
110kg (carriage)
16kg (trail eyes)
1.45kg (sight Type A)
1.57kg (sight Type B)
Length: 2.65m (travelling)
Barrel length: 1.94m (overall)
Width: 1.53m
Height: 1.05m
Elevation: +40° to +80°
Range: 8,300m (max)
400m (min)

This is a smooth bore mortar. When travelling it is carried complete with its baseplate on a two-wheeled carriage which can be towed behind a truck. For use in difficult terrain however it can be broken down into four components — barrel, bipod, baseplate and carriage, and then carried by pack mule. The mortar is suitable for mounting in an armoured vehicle, eg the M113 APC. The weapon is fired by pulling a lanyard which is attached to the firing lever. The baseplate is of all-welded construction and the socket is so designed to allow full 360° traverse without moving the baseplate. The bipod incorporates the recoil buffer with barrel clamping collar, traversing, elevating and divergence correction gears.

The following types of bomb have been developed for this mortar:

HE Bomb
This weighs 12.6kg with a Diehl fuze and contains a total of 2.25kg of TNT. The bomb is made of forged steel with the tail unit of light metal construction. The basic bomb is called the M58F and has a max range of 6,200m; a new mortar bomb has recently been introduced which has a max range of 8,300m. Israeli Military Industries has also developed 120mm mortar bomb with rocket assist which has a range of 10,500m.

Smoke Bombs
Two types of smoke bomb are available, $TiCl_4$ and Plastic Phosphorus. Ballistic data is identical to the HE bomb. The $TiCl_4$ produces an instant smoke screen for a short period whereas the Plastic Phosphorus is a dual purpose bomb. The Plastic Phosphorus takes a little longer to build up, but lasts longer.

The 120mm Soltam M-65 mortar in the firing position.

The 120mm Soltam M-65 mortar in the travelling position.

Practice Bomb
This is filled with an inert mass and incorporates a small smoke charge which is ignited by a regular fuze to mark the point of impact.

The propellant system is made up of a primary charge and a combination of up to nine secondary charges. The secondary charges are circular in shape and are positioned above the tail unit.

Employment
In service with the Israeli Defence Force and other undisclosed countries.

120mm Soltam Light Mortar

Israel

Calibre: 120mm
Weight: 120.57kg (firing)
218kg (travelling)
38.6kg (barrel)
5.4kg (breech piece)
37kg (bipod)
38kg (baseplate)
1.57kg (sight)
Barrel length: 1.73m (with breech piece)
Elevation: +45° to +80°
Range: 7,000m (max)
400m (min)

This mortar consists of four main components — barrel and breech piece, baseplate, bipod and sight. The three main components can be carried as individual loads by manpack or as mule packs. If required the complete mortar can be carried on a two wheeled carriage. This light mortar fires the same ammunition as the standard 120mm mortar. The breech is fitted with a fixed retractable striker, the striker can be retracted into the safe position by the turn of a safety lever. The bipod incorporates the recoil buffer, elevating, traversing, divergence correction gear and leg extending mechanism. The triangular baseplate is of all-welded steel construction and allows for full 360° traverse with out moving the baseplate.

The following types of mortar bomb are available:

HE Bomb M58F
This weighs 12.9kg complete with a Diehl fuze. It is

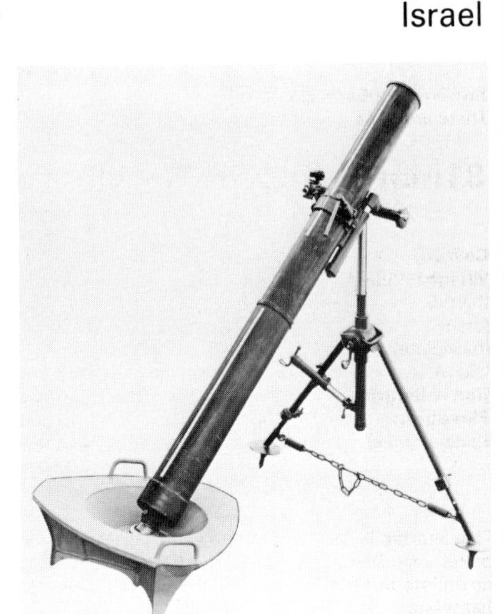

The 120mm Soltam light mortar in the firing position.

The 120mm Soltam light mortar in the travelling position.

of forged steel with the tail unit machined out of extruded light metal. The bomb is centred in the barrel by a driving band at its largest diameter. The M58F HE bomb has a max range of 6,200m and the new M58FF HE bomb has a max range of 7,000m.

Practice Bomb M58F
This is filled with an inert mass and incorporates a small smoke charge which is ignited by the regular fuze to mark the point of impact.

Smoke Bombs M58F
There are three types of smoke bomb:

$TiCl_4$ (FM)
Plastic White Phosphorus (PWP)
White Phosphorus (WP)
The propellant system is made up of primary cartridge, additional charge and eight secondary charges.

Employment
In service with the Israeli Defence Force and other undisclosed countries.

81mm Soltam M-64 Mortar

Israel

	Short Barrel	*Long Barrel*	*Type C*
Calibre:	81.4mm	81.4mm	81.4mm
Weight: (firing)	37kg	40kg	43kg
(barrel)	11.5kg	14.5kg	17.5kg
(bipod)	12.3kg	12.3kg	12.3kg
(baseplate)	13.2kg	13.2kg	13.2kg
(sight)	1.45 or 1.57kg	1.45 or 1.57kg	1.45 or 1.57kg
Barrel length:	1.155m	1.455m	1.455m
Elevation:	+43° to +80°	+43° to +80°	+43° to +80°
Range: (max)	4,000m	4,600m	4,600m
(min)	150m	150m	150m

This mortar is available in three models — short barrel, long barrel and Type C. These can be broken down into three loads consisting of barrel, bipod and baseplate.

The circular baseplate is of welded steel construction and the ball socket is so designed that full 360° traverse is available without moving the baseplate. The bipod incorporates the recoil buffer, elevating, traversing and divergence correction gears.

The Type C is a modified version of the long barrel model and differs from its parent model only in barrel design. The barrel consists of two parts of approximately equal length joined by an overlapping interrupted thread, ensuring a tight fit and eliminating any possible gas escape. Total length of the barrel and ballistic data is identical to the long barrel model.

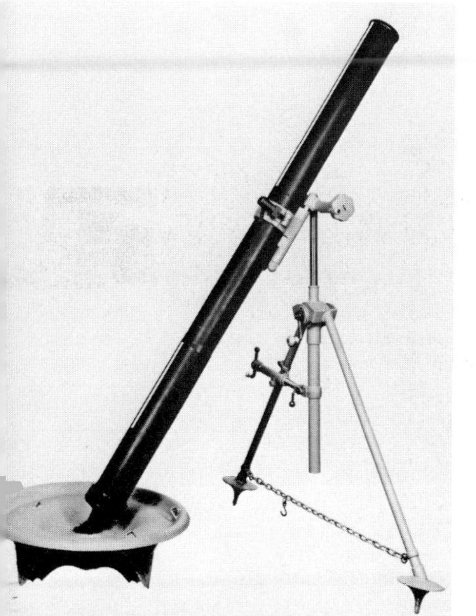

The standard long-barrelled 81mm mortar in the firing position.

The Type C 81mm mortar in the firing position.

The following types of mortar bomb are available for this weapon:

HE Bomb
This weighs 4.2kg complete, is of forged steel construction with a light metal tail unit.

Practice Bomb
This is similar to the other Soltam/Tampella practice bombs.

Smoke Bombs
There are two types — $TiCl_4$ and Plastic Phosphorus. The propellant system is made up of a primary charge and a combination of up to seven secondary charges. The secondary charges are circular in shape and are positioned above the tail unit.

Employment
In service with the Israeli Defence Force and other undisclosed countries.

Note: Soltam manufactures mortars made basically to Finnish Tampella designs, although in recent years Soltam has further developed most of the original Finnish mortars to meet specific Israeli requirements.

105mm Model 56 Pack Howitzer — Italy

Calibre: 105mm
Weight: 1,290kg
Length: 3.65m (travelling)
Barrel length: 1.478m (inc muzzle brake)
Width: 1.5m
Height: 1.93m (travelling)
Elevation: –5° to +65°
Traverse: 36° (total)
56° (total in anti-tank role)
Rate of fire: 4rpm for 30min
Range: 10,575m (HE projectile)
13,000m (RAP)
Ammunition: HE, Smoke, HEAT, Target indicating, Illuminating (7 charges)
Armour penetration: 102mm with HEAT projectile
Crew: 7
Towing vehicle: FIAT AR-59 (4×4); LWB Land Rover (4×4); Unimog (4×4)
Airportable: AB-205 helicopter — one load
Wessex helicopter — one load
C-130 aircraft can carry 2 guns and 2 towing vehicles plus crew

The 105mm Model 56 Pack Howitzer was developed by OTO Melara to meet the requirements of the Italian Army and entered production in 1957, since then over 2,000 have been built. The Model 56 Pack Howitzer is used by airborne, mountain and light artillery regiments where the weight of the weapon is the overriding factor rather than range. The weapon can be quickly disassembled into 11 components for transport by animal with the heaviest component weighing 122kg. The weapon fires the same range of 105mm semi-fixed ammunition as that used by the American 105mm M101 howitzer, in addition the Italian company of SNIA have developed an RAP with a max range of 13,000m.

A feature of the 105mm Model 56 Pack Howitzer is that it can also be used in the anti-tank role with a very low profile. In this role the wheels are moved

105mm Model 56 Pack Howitzer in the firing position.

from their normal position forward into sockets on the suspension arms. This lowers the profile of the weapon and, as the trails are much wider apart, enables the ordnance to be traversed 28° left and 28° right rather than the normal traverse of 18° left and 18° right.

When travelling the two rear sections of the trails fold up and the two main arms lock together. It is recognisable by its shield, multi-baffle muzzle brake and its folded trails whilst travelling. The German weapons have a much larger single-baffle muzzle brake.

Employment
Argentina, Australia, Bangladesh, Belgium, Canada, Chile, France, Ghana, Germany (FRG), India, Iran, Italy, Malaysia, New Zealand, Nigeria, Pakistan, Philippines, Saudi Arabia, Spain, Sri Lanka, Sudan Venezuela, United Arab Emirates, Yemen (South), Zambia and Zimbabwe.

Japanese Postwar Artillery

In the 1950s Japanese industry manufactured 57mm and 75mm recoilless rifles and 81mm mortars for United States units based on the Far East. In recent years Japan has manufactured the Swiss Oerlikon twin 35mm GDF-001 anti-aircraft guns as well as a 106mm recoilless rifle. The latter has the designation of the Type 60 and is in service on two self-propelled modes, first a single model mounted on the rear of a Jeep and second two mounted on the Type 60 self-propelled recoilless gun. Details of the latter are given in the companion volume *Armoured Fighting Vehicles of the World.*

Japanese Type 60 106mm recoilless rifles mounted on Jeeps, the ranging rifle is mounted over the top of the barrel at the rear.

20mm FK 20-2 Light AA Gun — Norway

Calibre: 20mm
Weight: 440kg (firing)
620kg (travelling)
Length: 4m (travelling)
Width: 1.86m (travelling)
Height: 2.2m (travelling)
Axis of bore: 0.58m (firing)
G/clearance: 0.38m
Elevation: −8° to +83°
Traverse: 360°
Range: 1,500-2,000m (effective AA)
Rate of fire: 800/900rpm (cyclic)
Crew: 1 (on gun)
Towing vehicle: Jeep or similar vehicle

The 20mm FK 20-2 light anti-aircraft gun was developed to meet the requirements of the West German and Norwegian armed forces for a light and highly mobile anti-aircraft gun. It basically consists of a modified Swiss HSS669 mount fitted with the German Rheinmetall 20mm Rh202 cannon, cradle, ammunition cases and flexible feed chutes designed by A/S Kongsberg Vapenfabrikk of Norway and a Kern optical sight. For transport the weapon is carried on a lightweight two-wheeled carriage and when in the firing position is supported on three trails.

The 20mm gun is provided with a 75-round magazine on either side with a further ten round magazine mounted over the top of the weapon, the latter normally contains AP ammunition. The following types of ammunition can be fired: APDS-T (m/v 1,150m/sec), API-T (m/v 1,100m/sec), HEI (m/v 1,045m/sec), HEI-T (m/v 1,045m/sec), TP/TP-T (m/v 1,045m/sec) and training break up ammunition.

Elevation and traverse is manual and the optical sight has a magnification of ×5 for use against ground targets and ×1 for engaging aerial targets.

Employment
FRG and Norway.

20mm FK 20-2 light AA gun of the German (FGR) Army in the firing position.

South African Artillery Development

For some years South Africa has been capable of manufacturing ammunition for the 17-pounder anti-tank guns, 25-pounder field guns (and the self-propelled Sexton version), 5.5in guns and the 105mm M7 self-propelled howitzers used by the South African Army.

A 90mm anti-tank gun is also in service with the South African Army, this is essentially a British 6-pounder anti-tank gun with the original 57mm ordnance replaced by a new 90mm one which may be based on that fitted to the Eland light armoured car. The latter is the Panhard AML-90 which was built under licence in South Africa until recently. The 90mm anti-tank gun fires a 90mm HEAT projectile which will penetrate 300mm of armour at a range of 500m.

More recently South Africa has developed a 155mm gun-howitzer under the designation of the G-5, this will replace the 25-pounder and 5.5in guns at present in use. The G-5 is mounted on a split trail carriage with four road wheels with walking beam suspension similar to that employed on the Israeli Soltam 155mm M-71 and M-68 gun-howitzers. When in the firing position the wheels are raised off the ground and the carriage is supported by a firing jack. An auxiliary propulsion system allows the G-5 to propel itself across the battlefield over short distances and the ordnance is fitted with a muzzle brake and a large fume extractor.

105mm/26 Howitzer — Spain

Calibre: 105mm
Weight: 1,950kg
Length: 5.86m (firing)
6.08 (travelling)
Barrel length: 3.349m
Width: 2.1m (travelling)
Height: 2.2m (travelling)
Track: 1.732m
Elevation: −5° to +45°
Traverse: 50° (total)
Range: 11,450m
Rate of fire: 4rd in first ½min
16rd in first 4min
Ammunition: Spanish projectile weight 15.27kg, m/v 504m/sec
US projectile weight 14.57kg

This was developed in the early 1950s from an earlier 1943 weapon which is still used as a training weapon as well as a saluting piece.

105mm/26 towed howitzers of the Spanish Army; trucks are M-34s.

Esperanza Mortars — Spain

	L	L1	L	L	SL
Calibre:	81mm	81mm	105mm	120mm	120mm
Weight (barrel)	19kg (long)	17kg	37.7kg	61kg	47kg
(baseplate)	13.5kg	13.5kg	43kg	100kg	43kg
(tripod)	10.5kg	11.5kg	24.25kg	40kg	31kg
(sight)	2kg	2kg	2kg	2kg	2kg
Barrel length:	1.45m (long) or	1.15m	1.5m	1.6m	1.6m
Range: (max)	4,500m (long) or	4,125m	6,000m	5,700m or 5,940m	5,000m or 5,940m
Rate of fire:	15rpm	15rpm	12rpm	12rpm	12rpm

The Esperanza range of mortars is manufactured by Esperanza Y. Cia, SA of Marquina, Vizcaya, Spain, and is known as the ECIA range for short.

81mm Mortar L and L1

The main differences between these two models is that the elevating and traversing screws are fully enclosed on the Model L1. For easy transportation the mortar can be broken down into three man-pack loads. There are two models, long barrel and standard barrel. The actual calibre of the mortar is 80.65mm. There are a total of six charges and the following mortar bombs are available: HE, high capacity, HE, smoke, illuminating and practice. Basic data of the HE rounds is given below, both have an effective damage radius of 100m.

	Type N	*Type NA*
Length:	381mm	354mm
Weight:	4.13kg	3.2kg
(filling)	675gr (TNT)	496gr (TNT)

81mm mortar Model L in the firing position.

The 81mm mortar model L has been replaced by the model L-N and the 81mm mortar model L1 has been replaced by the model L-L. The standard barrel model can fire an NA bomb to 4,125m or an N bomb to 4,270m while the long barrelled mortar can fire the NA bomb to 4,600m and an N bomb to 5,200m.

105mm Mortar Model L
This is a scaled up 81mm mortar and can be fired by gravity or trigger. Two types of carriage have been developed, one of these carries the barrel, baseplate and tripod, whilst the other model in addition carries 12 mortar bombs. This is normally towed by a 4×4 truck. Actual calibre of this mortar is 104.25mm and total weight travelling (including carriage) is 239kg. A full range of bombs is available including HE, smoke and practice. Basic data of the HE bomb is: effective damage radius 150m, length 546mm, weight 9.2kg, weight of TNT filling 1.704kg.

105mm mortar Model L in the firing position.

120mm mortar Model SL in the firing position.

120mm mortar Model L in the firing position.

120mm Mortar Model L
This is similar in construction to the 105mm mortar and is also carried on a two-wheeled carriage, total weight travelling (including carriage) is 328kg. There are a total of six charges and the actual calibre of the barrel is 119.6mm. Smoke, practice and HE bombs are available. Basic data of the HE bombs is:

	Type N	*Type L*
Length:	668mm	604mm
Weight:	16.745kg	13.195kg
(filling)	3.175kg	2.34kg
Max range:	5,700m	5,940m

120mm Mortar Model SL
This is a lightweight model of the above mortar and is carried on the lightweight carriage of the 105mm mortar (ie the model without the capacity for carrying mortar bombs), total weight is 257kg. It fires the same ammunition as the 120mm Mortar Model L. The model SL fires a N mortar bomb to a max range of 5,000m (four charges), or an L mortar bomb to 5,940m (five charges).

Employment
All of these mortars are in service with the Spanish Army.

155mm FH-77 Field Howitzer

Sweden

Calibre: 155mm
Weight: 11,000kg
Length: 10.86m at 0° elevation (firing)
11.59m (travelling)
Width: 7.18m (firing)
2.64m (travelling)
Height: 2.75m at 0° elevation (firing)
2.75m (travelling)
Elevation: −3° to +50°
Traverse: 25° left and right with elevation of less than +5°
30° left and right with elevation of more than +5°
Rate of fire: 3rd in 8sec
Range: 22,000m (charge 6)

The FH-77 has been developed by Bofors to meet a Swedish Army requirement for a new weapon with a high rate of fire, long range, and improved road and cross country mobility. A total of three prototypes were built and delivered to the Swedish Army by the end of 1973, and first production FH-77s were delivered in 1978.

For normal operations it is towed by the Saab-Scania SBAT (6×6) truck but it can however propel itself at a speed of up to 8km/h by its own integral power unit. This consists of a Volvo B20 which is connected to two variable hydraulic pumps which in turn are connected to hydraulic motors in the driving wheels.

Elevation and traverse is carried out hydraulically as is the ramming of the complete round. The new HE shell has been developed by Bofors and weighs 42.4kg complete with fuze. There are a total of six increments with a maximum m/v of 774m/sec. The plastic cartridge cases have a steel base and weigh 5.9kg without the charge. A supercharge is being developed which will increase muzzle velocity to 815m/sec and range to just over 23,000m. An RAP is being developed which will have a range of between 27,000 and 30,000m. Other projectiles that can be fired by the FH-77 include illuminating and smoke.

The FH-77 has a crew of six men. Each member of the gun crew has been provided with a headset which is dual purpose, first it acts as an ear defender and second as a communications headset. The commander and layer can speak to each other whilst the other crew members can only receive orders. In addition there is a direct link to the battery command post.

Variants
Under development is the FH-77B. This is being developed for the export market and will fire all NATO 155mm ammunition, have a screw type breech in place of the vertical sliding breech and have a maximum elevation of +70° compared to the basic FH-77 (or FH-77A as it is sometimes called) of +50°.

Employment
In service with the Swedish Army.

155mm Field Howitzer 77 in the firing position.

150mm m/39 Field Howitzer

Sweden

Calibre: 149.1mm
Weight: 5,720kg (travelling)
Length: 7.27m (firing)
6.55m (travelling, barrel to rear)
7.37m (travelling barrel forward)
Barrel length: 3.6m
Width: 5.53m (firing)
2.5m (travelling)
Height: 2.5m (travelling)
Elevation: −5° to +66°
Traverse: 45° (total)
Range: 14,600m
Ammunition: HE projectile weight 41.5kg, m/v 580m/sec, 8 charges
Rate of fire: 4-6rpm
Towing vehicle: Volvo TL31 L3153, 6×6, Truck

Whilst in the travelling position the barrel is withdrawn out of battery to the rear, chain drive being provided on the right hand side of the carriage for this purpose. The trails are of the split type, each trail being provided with a spade. There is also a castor wheel to assist in opening out the trails. The m/39 is fitted with old type wheels with solid rubber tyres whilst the m/39b has modern type wheels and rubber tyres.

When in action the weapon is very similar in appearance to the British 5.5in gun.

Employment
Sweden.

The m/39b. The barrel in this photograph has been drawn to the rear. The spades can be seen on the top of the trails, as can the castor wheel.

105mm 4140 Field Howitzer

Sweden

Calibre: 105mm
Weight: 2,800kg (firing)
3,000kg (travelling)
Length: 6.8m (travelling)
Barrel length: 3.36m (28 calibres)
Width: 1.81m (travelling)
Height: 1.85m (travelling)
Elevation: −5° to +60°
Traverse: 360°
Range: 15,600m
Rate of fire: 25rpm (max)
8rpm (normal)
Ammunition: HE, projectile weight 15.3kg m/v 640m/sec
Towing vehicle: Volvo TL31 L3153, (6×6) truck

105mm 4140 field howitzer in the firing position.

This weapon was developed by Bofors after the end of World War 2 and to some extent resembles the 10.5cm leichte Feldhaubitze 43 developed by Skoda of Czechoslovakia to meet the requirements of the German Army. This was not placed in production.

The 105mm 4140 field howitzer has four trails, when travelling the front two are locked together and raised upwards and locked in position under the ordnance while the rear two are locked together and attached to the towing vehicle. When in the firing position each of the four trails is staked to the ground and the weapon can then be quickly traversed through 360°.

Employment
Sweden.

Swedish Coastal Artillery

The Swedish Navy is responsible for coastal defence and has 25 mobile and 45 static coastal batteries which are armed with 75mm, 105mm, 120mm, 152mm and 210mm guns and Rb08 surface-to-air missiles.

90mm PV-1110 recoilless rifle mounted on a Haflinger light vehicle.

90mm PV-1110 Recoilless Rifle — Sweden

Calibre: 90mm
Weight: 260kg (total firing and travelling)
125kg (gun)
135kg (carriage)
Length: 3.691 (firing)
4.1m (travelling)
Barrel length: 3.691m
Width: 1.375m (travelling)
Height: 0.87m (travelling)
Elevation: −10° to +15°
Traverse: 75° to 110° (total) depending on the firing height
Range: 700m (effective)
Rate of Fire: 2rd in 13sec
6rd in 1min
Ammunition: Complete round weight 9.6kg
projectile weight 3.1kg, m/v 715m/sec, fin-stabilised, hollow charge
Armour penetration: 380mm at 90°, range 700m
Crew: 2-3

The 90mm PV-1110 (PV standing for pvpjäs) was built by Bofors in Sweden. An interesting feature of the weapon is that its overall height can be adjusted from 0.87 to 0.47m. The weapon is aimed by a 7.62mm spotting rifle (10 rounds for ready use) mounted over the barrel, the sight is to the left of the barrel, the weapon is fired with the aid of a pistol type grip under the barrel.

The weapon can also be mounted on a sled, a Volvo (4×4) Laplander or an Austrian Haflinger vehicle, it can also be towed by the latter vehicles. When used in the Swedish infantry brigades it is normally towed and when with the Swedish armoured brigades it is mounted on the Volvo Laplander vehicle.

Employment
Ireland and Sweden.

57mm m/54 AA Gun

Sweden

Calibre: 57mm
Weight: 8,100kg (travelling)
Length: 7.27m (firing)
8.435m (travelling)
Barrel length: 3.42m
Width: 3.8m (firing)
2.38m (travelling)
Height: 2.435m (firing)
2.81m (travelling)
Elevation: −5° to +90°
Traverse: 360°
Rate of fire: 160rpm (cyclic)
Ammunition: Projectile weight 2.6kg, m/v 920m/sec
Range: 4,000m (effective AA)
14,000m (max horizontal)
Towing vehicle: Volvo TL31 L3154 (6×6) truck

This was built in Sweden by Bofors and is mounted on a four-wheeled carriage; when in the firing position the wheels are kept on and the carriage rests on four jacks, one at each end of the carriage and another either side mounted on stabilisers that swing out from the side of the carriage.

The weapon is easily recognisable by its very large shield which is flat at the front and has flat side pieces to it. Ready use ammunition racks are provided at the rear of the platform. As with all of the Bofors AA guns the barrel is provided with a suppressor.

Employment
Belgium, Sweden.

The m/54 57mm AA gun in the firing position.

40mm m/48 (L/70) Light AA Gun

Sweden

Calibre: 40mm
Weight: 4,800kg (travelling)
Length: 7.29m (overall travelling)
Barrel length: 2.8m
Width: 2.25m (travelling)
Height: 2.349m (travelling)
Elevation: −4° to +90°
Traverse: 360°
Rate of fire: 240rpm (cyclic)
Range: 3,000m (effective AA)
Crew: 6 (2 on gun; 4 handling/loading ammunition)
Towing vehicle: Volvo TL22 L2204 (6×6) truck (Sweden)

This gun was first developed after the end of World War 2, the first prototype being built in 1947. In 1951 it entered production and since then it has been exported all over the world.

The carriage has racks for a total of 48 rounds of ready use ammunition, the ammunition is fed into guideways above the breech, a total of 16 rounds can be held above the breech for immediate use. The empty cartridge cases are ejected via a chute in the lower half of the shield. The weapon is mounted on a four-wheel carriage. When in firing position the weapon rests on four jacks, one at each end of the carriage and the other two on stabilisers that swing

Bofors 40mm L/70 AA gun Model A with integral generator mounted at rear of carriage.

out either side of the carriage.

There are two basic models of the Bofors L/70, the Model A which has its own integral generator and the Model B which requires an external power source when being used in the powered mode. When in the latter mode max elevation speed is 45°/sec and max traverse speed is 85°/sec. A wide range of fixed ammunition is fired, this being fed to the gun in clips of four rounds. The following types can be fired: AP projectile, weight 0.93kg (m/v 1,025m/sec), pre-fragmented projectile, weight 0.88kg (m/v 1,025m/sec), and HE and TP-T projectiles, weight 0.96kg (m/v 1,005m/sec).

The L/70 can be used in the optical mode but is normally used in conjunction with a fire control system such as the Dutch L4/5 or the more recent Flycatcher, or the Swiss Skyguard or Super-Fledermaus systems, all of which have their own entries in the artillery fire control systems section of *Artillery of the World*.

One of the more recent developments is the Bofors BOFI gun system which consists of the Bofors L/70, BOFI (Bofors Optronic Fire-control Instrument) and the new proximity fuzed ammunition which enables targets to be engaged under both day and night conditions with a high probability of a target kill.

Employment

Argentina, Austria, Belgium, Denmark, Dijbouti, France, West Germany, India (manufactured under licence), Iran, Italy (manufactured under licence), Israel, Ivory Coast, Libya, Malaysia, Netherlands, Nigeria, Norway, Peru, South Africa (manufactured under licence), Uganda, UK and Venezuela.

Bofors 40mm L/70 AA gun Model A with BOFI equipment and radar.

40mm m/36 Light AA Gun

Sweden

Calibre: 40mm
Weight: 2,150kg (m/38)
2,400kg (m/39)
2,050kg (m/48E)
Length: 6.3m (m/38) (travelling)
6.38m (m/39)
6.39m (m/48E)
Barrel length: 2.24m (56 cal)
Width (firing): 3.98m (m/38)
3.92m (m/39)
3.92m (m/48E)
Width: 1.72m (travelling all models)
Elevation: −5° to +90°
Traverse: 360°
Rate of fire: 120/140rpm (cyclic)
70rpm (practical)
Ammunition: AP projectile weight 0.89kg, m/v 850m/sec
HE projectile weight 0.955kg, m/v 850m/sec
Range: 2,560m (effective AA)

The 40mm m/36 was the original Bofors anti-aircraft gun which was introduced into service with many armies shortly before the start of World War 2. The gun was also produced under licence in Austria, Belgium, Finland, France, Greece, Hungary, Italy, Norway, Poland and the UK. During World War 2 new versions were produced in Canada, the UK and the USA, these were to a new design and were quicker and cheaper to manufacture.

When in the firing position, the carriage is supported on four screw jacks, one at either end of the carriage and one either side on outriggers. There are three slightly different carriages, these being called the m/38, m/39 and m/48E.

Employment
Argentina, Brazil, Colombia, Finland, Ireland, Nepal, Sweden, Zaire and other countries.

Bofors 40mm m/36 light AA gun in the firing position.

120mm m/41C (1941) Mortar

Sweden

Calibre: 120.25mm
Weight: 285kg (firing)
600kg (travelling)
Barrel length: 2m
Elevation: +45° to +80°
Traverse: 360°
Range: 6,400m
Rate of fire: 12-15rpm
Ammunition: HE shell weight 13.3kg, m/v 125-317m/sec, plus other types of bomb

This is the Finnish Tampella M-1940 120mm mortar produced in Sweden. In 1956 the M-56 stand (produced under licence from Hotchkiss Brandt of France) replaced the earlier Tampella-type stand. Eight m/41Cs are in the mortar company of the infantry brigades and six m/41Cs in the mortar company of the Norrlands Brigades. In 1972 the m/41Cs were fitted with new sights.

Employment
Ireland and Sweden.

120mm m/41C mortar.

81mm m/29 (1929) Mortar

Sweden

Calibre: 81.4mm
Weight: 60kg (firing)
Barrel length: 1m
Elevation: +45° to +80°
Traverse: 90° (total)
Range: 2,600m (max)
Rate of fire: 15-18rpm
Ammunition: HE, shell weight 3.5kg, m/v 70-190m/sec
Crew: 2-3

This is the French Stokes-Brandt 81mm M-1917 Mortar produced in Sweden, it was built until 1943. In 1934 Swedish sights replaced the French sights. Six m/29s are in the mortar platoon of the heavy weapons company of each Swedish infantry battalion.

The m/29 is very similar in appearance to the American 81mm mortar M1, they both have a rectangular base with a carrying handle, bipod with chains between them and the left leg has an anti-cant device fitted to it.

Employment
Ireland and Sweden.

m/29 81mm mortar in the firing position with its ammunition.

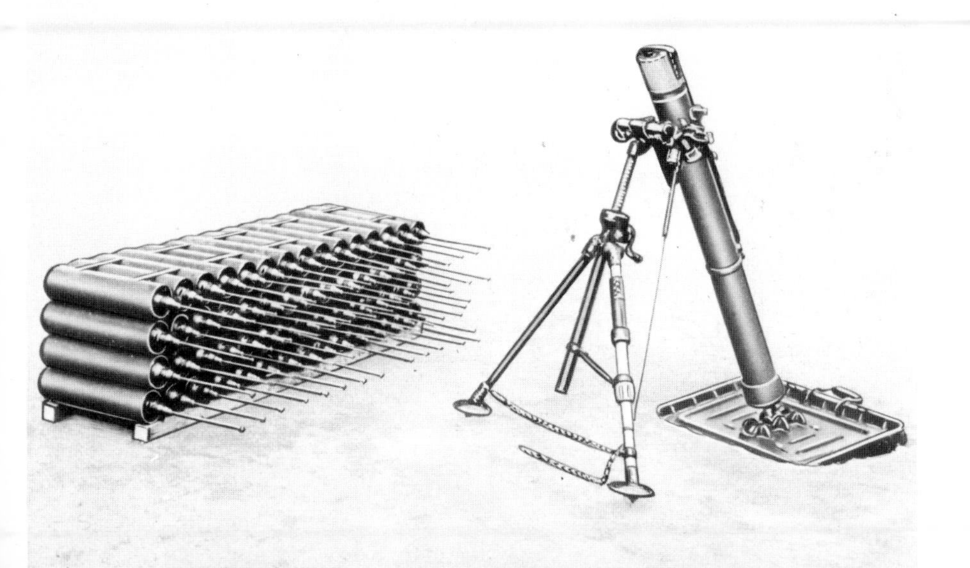

Swiss Artillery

9cm Panzerabwehrkanone 1950
This light anti-tank gun is known as the 9cm Pak 50 for short. It has a lightweight split trail type carriage and a square shield, the barrel is fitted with a muzzle brake.

9cm Panzerabwehrkanone 1957
This is almost identical to the above model but has a circular shield and is known as the Pak 57 for short. It has a longer range than the earlier model and fires improved ammunition.

10.5cm Haubitze 1946 L22
This is known as the 10.5cm Hb 46. Its carriage is of the split trail type and the weapon has an elevation from +0° to +65°, total traverse being 60°. It fires a HE projectile weighing 15.5kg with a max m/v of 490m/sec.

10.5cm Kanone 1935 L42
This is known as the 10.5cm Kan 35 for short. When it first entered service it had large steel wheels with solid rubber tyres and a two wheeled limber, again with solid tyres. More recently it has been fitted with new wheels and conventional tyres. The weapon has an elevation of +45° and a depression of −3°, total traverse being 60°.

Note: The 15cm Haubitze 1942 L28 (howitzer) and the 34mm Flab Kanone 38 (anti-aircraft gun) have now been withdrawn from service with the Swiss Army.

Weapon	Weight	Range	ROF
9cm Pak 50	631kg	600m	20rpm
9cm Pak 57	716kg	800m	20rpm
10.5cm Hb 46	1,800kg	12,000m	10rpm
10.5cm Kan 35 L42	4,245kg	21,000m	5rpm

Employment
All in service with Swiss Army, inc reserve elements.

The Swiss 9cm Pak 57 anti-tank gun.

The Swiss 10.5cm field gun M-35 as in service today.

The Swiss 10.5cm field howitzer M-46 in the firing position.

81mm Model 1933 Mortar

Switzerland

Calibre: 81mm
Weight: 62kg (total)
21kg (barrel)
18kg (mount)
21kg (baseplate)
2kg (sight)
Barrel length: 1.265m
Elevation: +45° to +85°
Traverse: 8°
Crew: 3

The 81mm Mortar Model 1933 is called the 8.1cm Minenwerfer Model 1933 or 8.1cm Mw 33 by the Swiss Army. It is of Swiss design and construction and is similar to most mortars of the prewar period. The rectangular baseplate is provided with a carrying handle. The two legs are connected by a chain with the cross levelling gear being connected to the left leg of the tripod. The sight is on the left side.

The Model 1933 mortar fires the same mortar bombs as the Model 1972 described below to the same ranges.

81mm mortar Model 1933 in the firing position.

81mm Model 1972 Mortar

Switzerland

Calibre: 81mm
Weight: 45.5kg (total)
12kg (barrel)
15kg (mount)
16.5kg (baseplate)
2kg (sight)
Barrel length: 1.267m
Elevation: +45° to +90°
Traverse: 10°
Crew: 3

This 81mm mortar is the replacement for the earlier Model 1933 mortar. Its full designation is 8.1cm Minenwerfer 1972 or 8.1cm Mw 72 for short. Like the earlier mortar this mortar can be broken down into three man pack loads for easy transportation. The baseplate consists of a circular flat disc with the webs welded to the underside. The traversing gear is on the right side of the bipod with the sight on the left. The following types of bomb are fired by this weapon:

	HE	*Smoke*	*Fragmentation*
Weight:	3.17kg	3.67kg	6.89kg
No of charges:	0-6	0-6	0-6
M/v: (min)	70m/sec	70m/sec	64m/sec
(max)	260m/sec	260m/sec	110m/sec
Range: (max)	4,100m	3,000m	1,070m

The new 81mm mortar Model 1972.

120mm Model 1974 Mortar — Switzerland

Calibre: 120mm
Weight: 620kg (total)
87kg (barrel)
95kg (baseplate)
Barrel length: 1.77m
Elevation: +45° to +85°
Traverse: 10° (360° when mounted in APC)
Crew: 4

This is known as the 12cm Minenwerfer 1964 or 12cm Mw 64 for short. It is in service in two models. One is mounted in the rear of a modified M113A1 APC, this being known as the MvPz64, if required the mortar can be taken out of the vehicle and mounted on a baseplate which is carried on the side of the vehicle. The second model is mounted on a two wheeled carriage which can be towed by a jeep, in addition this carriage carries 12 ready use mortar bombs.

The following types of mortar bomb are used with this mortar:

	HE	*Smoke*
Weight:	14.33kg	14.33kg
No of Charges:	1-8	1-8
M/v: (min)	128m/sec	128m/sec
(max)	420m/sec	420m/sec
Range: (max)	9,300m	9,300m

Employment
All of these mortars are in service only with the Swiss Army.

35mm Oerlikon GDF-001 Twin AA Gun — Switzerland

This was formerly known as the 2 ZLa/353MK

Calibre: 35mm
Weight: 6,700kg (travelling and firing with ammunition)
6,300kg (travelling and firing w/o ammunition)
Length: 7.8m (travelling)
Barrel length: 3.15m (90 calibre)
Width: 2.26m (travelling)
Height: 2.6m (travelling)
1.72m (emplaced)
Elevation: −5° to +92°
Traverse: 360°
Range: up to 4,000m (tactical)
Rate of fire: 550rpm/barrel, cyclic
Ammunition supply: One feed-hopper per barrel with 56rd, in addition one reloading hopper per weapon with 63rd each
Crew: 3 (on gun)
Towing vehicle: 5ton 6×6 or 4×4 truck

The GDF-001 is the largest calibre gun in the Oerlikon range and is used in conjunction with the Contraves fully automatic fire control system Super-Fledermaus or Skyguard. The electro-hydraulic system on the gun mounting enables the three man crew to change over from the travelling to fire position, including levelling, in only 1½min. In the firing position the wheels are lifted clear of the gound and the gun rests on three stabilisers. Elevation speed is 56° a second and traverse speed is 112° a second. Ammunition supply to the guns is based on a fully automatic clip-feed conveyor system. Each gun is provided with a two wheeled power supply unit. Oerlikon have recently introduced a modification package for the GDF-001 called the NDF-A. This consists of a full camouflage kit which is available for use immediately after taking position, improved maintenance, integrated automatic breech lubrication system and a Ferranti type GSA Mk 3 auxiliary sight for use in both anti-aircraft and ground engagements. Ammunition is fed to each gun in seven-round clips and the following types of fixed ammunition, all with a muzzle velocity of

1,175m/sec can be fired: Mine HEI-T with projectile weight 0.535kg, Mine HEI with projectile weight 0.550kg, SAPHEI-T with projectile weight 0.55kg, TP-T with projectile weight 0.55kg and TP with projectile weight 0.55kg.

Employment
Austria (known as M-65), Finland, Japan (manufactured under licence), South Afrcia, Spain, Switzerland (known by the Swiss Army as the K-63) and some other countries. A naval version of the GDF is known as the GDM-A.

The 35mm Oerlikon twin AA gun GDF-001 deployed for action.

30mm Oerlikon GCF-BM2 Twin AA Gun — Switzerland

Calibre: 30mm
Weight: 5,499kg (firing)
5,492kg (travelling)
Length: 7.47m (travelling)
Barrel length: 2.555m
Width: 2.40m (travelling)
Height: 2.435m (travelling)
Track: 2.057m
Elevation: −15° to +80°
Traverse: 360°
Range: 3,000m (effective AA)
Rate of fire: 650rpm/barrel (cyclic)
Crew: 1 (on gun)
Towing vehicle: 3,000kg (4×4) truck

The GCF-BM2 anti-aircraft gun was developed by the British Manufacture and Research Company of Grantham, England, to meet the requirements of a country in the Middle East, reported to be Saudi Arabia. It consists of a Rubery Owen FV2505 four-wheeled trailer fitted with an Oerlikon twin 30mm GCM mount. The trailer has its own integral generator and when being used in the powered mode the mount has a maximum traverse speed of 90°/sec and a maximum elevation speed of 60°/sec. When in the firing position the trailer is supported on four stabilisers which are lowered to the front, sides and rear.

Each of the 30mm KCB guns has 160 rounds of belted ammunition for ready use and the following types of ammunition can be fired: Mine HEI or Mine HEI-T (m/v 1,080m/sec), SAPHEI (m/v 1,080m/sec) and TP or TP-T (m/v 1,080m/sec), with all rounds weighing 0.87kg complete.

Employment
Reported to be in service with Saudi Arabia.

30mm Oerlikon twin AA gun GCF-BM2 in travelling order.

25mm Oerlikon GBI-AO1 AA Gun — Switzerland

Calibre: 25mm
Weight: 600kg (travelling w/o ammunition)
440kg (firing w/o ammunition)
Length: 4.72m (travelling)
Barrel length: 2.182m
Width: 1.8m (travelling)
Height: 1.65m (travelling)
G/clearance: 0.4m
Axis of fire: 0.5m (firing)
Elevation: −10° to +70°
Traverse: 360°
Range: 2,000m (effective AA)
Rate of fire: 570rpm (cyclic)
160rpm (practical)
Crew: 3 (1 on gun)
Towing vehicle: 4×4 light truck

The 25mm GBI-AOI is the latest Oerlikon light anti-aircraft gun and can also be used in the ground role. It is carried on a two-wheeled trailer and can be quickly brought into action by its three-man crew. When in the firing position the weapon is supported on its three trail legs.

The 25mm cannon used in the system is the KBA-C which is fed from two ammunition boxes each holding 40 rounds of belted ammunition, the gunner can select either single shots or full automatic fire. The following types of fixed ammunition can be fired: APDS-T, projectile weight 128g (m/v 1,360m/sec, this will penetrate 25mm of armour at an incidence of 0° at a range of 1,000m); HEI-T/HEI, projectile weight 180g (m/v 1,100m/sec); SAPHEI-T/SAPHEI, projectile weight 180g (m/v 1,100m/sec); and TP-T/TP, projectile weight 180g (m/v 1,100m/sec). The 25mm KBA series cannon is also installed in the Oerlikon GBD range of turrets which can be mounted on light AFVs.

Employment
Production. In service with undisclosed countries.

25mm Oerlikon GBI-AO1 AA gun in the firing position.

20mm Oerlikon GAI-DO1 AA Gun

Switzerland

Calibre: 20mm
Weight: 1,330kg (firing with ammunition)
1,800kg (travelling with ammunition)
Length: 4.59m (travelling)
Barrel length: 1.906m
Width: 1.86m (travelling)
Height: 2.34m (travelling)
Axis of fire: 0.6m (firing)
Elevation: −3° to +81°
Traverse: 360°
Range: 2,000m (effective AA)
Rate of fire: 1,050rpm/barrel (cyclic)
Crew: 4 (1 on gun)
Towing vehicle: 4×4 light truck

This can be used against both ground and aerial targets and is provided with a P56 Galileo electronic fire control computer, optical reflex sight with independent aiming line for anti-aircraft use and with a special ground target sight. Elevation and traverse is hydraulic with manual controls being provided for emergency use, maximum traverse speed is 80°/sec and maximum elevation speed is 48°/sec. The gunner can select single shots, rapid single shots, limited bursts or full automatic fire. Each cannon has 120 rounds of ready use ammunition and the following types of ammunition can be fired: AP-T. (m/v 1,150m/sec will penetrate 15mm of armour at an incidence of 0° at a range of 800m), Mine HEI/Mine HEI-T (m/v 1,100m/sec), SAPHEI-T/SAPHEI (m/v 1,100m/sec) and TP-T/TP (m/v 1,100m/sec); all of these projectiles weigh 125g except AP-T which weighs 110g.

If required the GAI-DO1 can be connected to a remote input equipment that provides target speed, crossover point range, when the target is within the effective range of the weapon. The guns can also be connected to a radar monitoring system such as the LPD20 and an alternative sight with a radar data recorder is also available.

Employment

In service with undisclosed countries.

The Oerlikon GAI-DO1 gun deployed.

20mm Oerlikon GAI-CO1 and GAI-CO4 AA Gun

Switzerland

Calibre: 20mm
Weight: 534kg (travelling with ammunition)
370kg (firing with ammuntion)
Length: 3.87m (firing)
Barrel length: 1.84m
Width: 1.7m (firing)
Height: 1.45m (firing)
Axis of bore: 0.5m (firing)
Elevation: −7° to +83°
Traverse: 360°
Range: 2,000m (effective AA)
Rate of fire: 1,050rpm (cyclic)
Crew: 3 (1 on gun)
Towing vehicle: 4×4 light truck

The Oerlikon GAI-CO1 was formerly known as the HS 693-B3.1 and can be used both in the anti-aircraft and ground-to-ground roles. Elevation and traverse is manual with the former being controlled by a handwheel and the latter by a foot pedal. The sight has a magnification of ×1 for engaging aerial targets and ×2.5 for engaging ground targets. The gunner can select either single shots or full automatic fire and ammuntion is fed to the KAD-B13-3 gun via a 75-round box magazine mounted on the right side. The following types of fixed ammuntion can be fired: AP-T, Mine HEI-T and Mine HEI, SAPHEI-T and SAPHEI, TP-T and TP, all of these projectiles weigh 125g and have a m/v of 1,100m/sec with the exception of the AP-T projectile which weighs 110g and has a m/v of 1,150m/sec, this will penetrate 15mm of armour at an incidence of 0° at a range of 800m.

The Oerlikon GAI-CO4 is almost identical to the GAI-CO1 but has an Oerlikon KAD-B14 dual feed cannon with two ammuntion boxes each of which contain 75 rounds of ammuntion.

Employment

Known users of the GAI-CO1 include Chile and South Africa. The basic KAD(HS 820) cannon was also manufactured in the USA for the M114A1 reconnaissance vehicle under the designation of the M139.

20mm Oerlikon GAI-CO1 AA gun in the firing position.

20mm Oerlikon GAI-CO3 AA Gun — Switzerland

Calibre: 20mm
Weight: 510kg (travelling with ammunition)
342kg (firing with ammunition)
Length: 4.27m (firing)
Barrel length: 2.24m
Width: 1.7m (firing)
Axis of bore: 0.5m (firing)
Elevation: −7° to +83°
Traverse: 360°
Range: 2,000m (effective AA)
Rate of fire: 1,050rpm (cyclic)
Crew: 3 (1 on gun)
Towing vehicle: 4×4 light truck

The 20mm Oerlikon GAI-CO3 was formerly known as the HS639-B4.1 and has the same carriage as the 20mm GAI-CO1 and GAI-CO4 automatic anti-aircraft guns. The gunner can select either single shots or full automatic fire and the 20mm KAD-AO1 automatic cannon (former designation HS820 SAA1) is fed from a 50-round drum magazine.

Elevation and traverse is manual with the anti-aircraft sight having a magnification of ×1 and the telescopic ground-ground sight having a magnification of ×2.5. The following types of fixed ammunition can be fired: AP-T, Mine HEI-T, Mine HEI, SAPHEI-T, SAPHEI, TP and TP-T. All projectiles weigh 125g and have a m/v of 1,100m/sec with the exception of the AP-T projectile which weighs 110g and has a m/v of 1,050m/sec, this will penetrate 15mm of armour at an incidence of 0° at a range of 800m.

Employment

In service with undisclosed countries.

20mm Oerlikon GAI-CO3 AA gun in the firing position.

20mm Oerlikon GAI-BO1 AA Gun — Switzerland

Calibre: 20mm
Weight: 547kg (travelling)
405kg (firing)
Length: 3.85m (travelling)
Barrel length: 2.4m
Width: 1.55m (travelling)
Height: 2.5m (travelling)
Axis of bore: 0.425m (firing)
Elevation: −5° to +85°
Traverse: 360°
Range: 2,000m (effective AA)
Rate of fire: 1,000rpm (cyclic)
Crew: 3 (1 on gun)
Towing vehicle: 4×4 light truck

The 20mm Oerlikon GAI-BO1 was formerly known as the Hispano-Suiza 10 ILa/5TG and is the lightest member of the extensive Oerlikon range of 20mm anti-aircraft guns which can also be used in the ground-ground role. For ease of transportation in rough terrain the mount and carriage can be disassembled into individual loads.

The gas operated cannon is designated the KAB-001 and is fed from one of three different types of magazine, 50-round drum, 20-round drum or an eight-round box. The following types of fixed ammuntion can be fired: AP-T, Mine HEI-T and Mine HEI, SAPHEI-T and SAPHEI, TP and TP-T. All projectiles weigh 125g and have a m/v of 1,100m/sec with the exception of the AP-T projectile which weighs 110g and has a m/v of 1,150m/sec and will penetrate 15mm of armour at an incidence of 0° at a range of 800m.

The Austrian Army uses a number of these weapons mounted on the rear of Pinzgauer (6×6) 1,500kg cross-country vehicles.

Employment

Known users include Austria, South Africa, Spain and Switzerland.

20mm Oerlikon GAI-BO1 AA gun in the firing position.

20mm Oerlikon GAI-BO1 AA gun mounted on rear of a Pinzgauer.

Hispano Suiza AA Guns

Switzerland

In 1972 Machine Tool Works Oerlikon-Bührle took over the Swiss company of Hispano Suiza who also manufactured an extensive range of automatic weapons and anti-aircraft guns, some of these were integrated into the Oerlikon range while others were phased out of production. Some of the more important anti-aircraft guns are listed below:

HS630 Full details of this mount with its three HSS 804 20mm cannon are given in the entry for the 20mm M-55 anti-aircraft gun under Yugoslavia.

HS639-B3.1 Now known as the Oerlikon GAI-B01 for which there is a separate entry.

HS639-B4.1 Now known as the Oerlikon GAI-C03 for which there is a separate entry.

HS639-B5 Now known as the Oerlikon GAI-C04 for which there is a separate entry.

HS661 Armed with a single HS831 30mm cannon.

HS666 Now known as the Oerlikon GAI-D01 for which there is a separate entry.

HS669 Armed with single 20mm cannon, no longer available.

HS673 Armed with a single 20mm cannon, no longer available.

180mm S-23 Gun

USSR

Calibre: 180mm
Weight: 20,400kg (travelling)
Length: 10.485m (travelling)
Barrel length: 8.8m
Width: 2.996m (travelling)
Height: 2.621m (travelling)
G/clearance: 0.4m
Elevation: −2° to +50°
Traverse: 44° (total)
Range: 30,400m (max normal ammuntion)
43,800m (max RAP)
Rate of fire: 2rpm (sustained)
1rpm (short period)
Ammunition: HE, projectile weight 88kg (m/v 790m/sec) concrete piercing and nuclear
Crew: 16
Towing vehicle: AT-T heavy tracked artillery tractor

The 180mm gun S-23 was developed in the 1950s and for many years was known in the west as the 203mm gun-howitzer M1955. In 1973 Israel captured a number of these weapons from the Syrian army and it was then found that the actual calibre of the weapon was 180mm. The ordnance is mounted on a split trail type carriage which has four rubber-tyred road wheels; when travelling the ordnance is withdrawn out of battery to the rear and secured over the closed trails, a two wheeled limber is attached to the ends of the trails for travelling. The S-23 has a pepperpot muzzle brake, screw breech mechanism and fires bag-type, variable charge separate loading ammuntion.

Employment

Egypt, India, Syria and the USSR.

180mm S-23s being towed by AT-T heavy tracked artillery tractors through Red Square, Moscow.

152mm D-20 Gun-Howitzer — USSR

Calibre: 152.4mm
Weight: 5,500kg (firing)
Length: 8.69m (travelling)
Barrel length: 5.055m
Width: 2.35m (travelling)
Height: 2.464m (travelling)
G/clearance: 0.4m
Elevation: −5° to +45°
Traverse: 58°
Range: 18,500m (HE round max)
Rate of fire: 5-6rpm
Ammunition: (separate loading) HE, projectile weight 43.6kg, m/v 655m/sec
APHE, projectile weight 48.8kg, m/v 600m/sec
Armour penetration: 124mm at 1,000m (APHE round)
Crew: 10
Towing vehicle: AT-L or AT-S tracked artillery tractor
Ural-375, 6×6 Truck

Tatra 813 (8×8) truck towing 152mm gun-howitzer D-20.

The D-20 uses the same carriage as the 122mm D-74 field gun and has replaced the 152mm M1937 (ML-20) Gun-Howitzer in most units.

The weapon has a semi-automatic vertical sliding wedge type breechblock and fires case-type, variable-charge, separate-loading ammunition. In addition to the HE and APHE rounds it also fires chemical, incendiary and illuminating projectiles.

The main difference between the D-20 and the D-74 is that the D-20 has a shorter, thicker stepped barrel with a muzzle brake. Other features are the same as the D-74: recoil system above the barrel, irregular shaped sloping shield with a vertical sliding centre section, the shield overlaps the wheels. The weapon can be quickly traversed through 360° by means of a firing jack that folds up under the barrel, it has split box section trails each with a caster wheel, this folds up on top of the trail when the weapon is being towed. Two folding type spades are also provided.

Employment
Bulgaria, Czechoslovakia, China (Type 66 but based on chassis of the 122mm Gun Type 60 which is copy of Soviet D-74), East Germany, Hungary and USSR; and other countries.

152mm M1943 (D-1) Howitzer — USSR

Calibre: 152.4mm
Weight: 3,600kg (firing)
Length: 7.558m (travelling)
Barrel length: 4.207m
Width: 1.994m (travelling)
Height: 1.854m (travelling)
Track: 1.8m
G/clearance: 0.37m
Elevation: –3° to +63.5°
Traverse: 35° (total)
Range: 12,400m (HE round max)
Rate of fire: 4rpm
Ammunition: HE, projectile weight 39.9kg, m/v 508m/sec
Semi-AP, projectile weight 51.1kg, m/v 432m/sec
A HEAT round is also available
Armour penetration: 82mm at 1,000m (Semi-AP round)
Crew: 7
Towing vehicle: AT-S medium tracked artillery tractor
Ural-375 (6×6) truck
ZIL-151 (6×6) truck
ZIL-157 (6×6) truck

This weapon has a screw type breechblock and fires case-type, variable-charge, separate loading ammunition. It has the same carriage and shield as the M1938 (M-30) 122mm howitzer.

It is recognisable by its short barrel with a double baffle muzzle brake, recoil system above and below the barrel (hydraulic buffer above and hydro-pneumatic recuperator below barrel), the centre section of the shield is raised when in the firing position, top half of the shield slopes to the rear. The carriage has two large single-tyred wheels and the trails are of a box section with folding spades.

Employment
Afghanistan, Albania, China, Germany (GDR), Egypt, Hungary, Ethiopia, Mongolia, Mozambique, Poland, Syria, USSR and Vietnam.

152mm howitzers M1943 (D-1).

152mm M1937 (ML-20) Gun-Howitzer USSR

Calibre: 152.4mm
Weight: 7,261kg (firing)
8,073kg (travelling)
Length: 7.21m (travelling excl limber)
Barrel length: 4.925m (inc muzzle brake)
Width: 2.312m (travelling)
Height: 2.26m (travelling)
Track: 1.9m
G/clearance: 0.315m
Elevation: −2° to +65°
Traverse: 58° (total)
Range: 17,265m (max)
Rate of fire: 4rpm
Ammunition: HE, projectile weight 43.6kg, m/v 655m/sec
APHE, projectile weight 48.8kg, m/v 600m/sec
Armour penetration: 124mm at 1,000m (APHE round)
Crew: 9
Towing vehicle: AT-S medium tracked artillery tractor

This weapon was used during World War 2 and has now been replaced by the D-20 in the Soviet Army although it is still in use with reserve units as well as members of the Warsaw Pact Forces. It uses the same carriage as the 122mm M1931/37 corps gun. The M1937 has a screw type breechblock and fires case-type, variable charge, separate loading ammunition, a modified version of this gun, the M1937/43 (ML-20S), is used in the ISU-152 assault gun.

It is recognisable by its twin road wheels, long barrel with a multi-baffle muzzle brake, the recoil system is under the barrel, whilst travelling the barrel is pulled back over the trails, small shield behind the vertical balancing gears, split section trails of a box-section with detachable picket spades. A limber is attached to the rear of the weapon when it is being towed.

Employment
Albania, Algeria, Bulgaria, China, Cuba, Czechoslovakia, Egypt, Finland, Germany (GDR), Hungary, Iraq, North Korea, Poland, Romania, Syria, Vietnam, Yugoslavia.

152mm gun-howitzer M1937 (ML-20) with its limber attached.

152mm M1938 (M-10) Howitzer USSR

Calibre: 152.4mm
Weight: 4,150kg (firing)
4,550kg (travelling)
Length: 6.399m (travelling)
Barrel length: 3.7m
Width: 2.096m (travelling)
Height: 1.9m (travelling)
Track: 1.655m
G/clearance: 0.305m
Elevation: −1° to +65°
Traverse: 50° (total)
Range: 12,400m (max)
Rate of fire: 4rpm
Ammunition: HE, projectile weight 39.9kg, m/v 508m/sec
Semi-AP, projectile weight 51.1kg, m/v 432m/sec
Armour penetration: 82mm at 1,000m (semi-AP round)
Crew: 7
Towing vehicle: AT-S medium tracked artillery tractor

This weapon was developed before World War 2, the M1940 (M-60) 107mm Gun which appeared at the same time used the same carriage as the 152mm M1938 (M-10). However, during the war the gun was mounted on the carriage of the M1938 (M-30) suitably modified, it was then known as the M1943 (D-1), see separate entry for this weapon.

The M1938 (M-10) fires case-type, variable-charge, separate loading ammunition and has a screw-type breechblock.

It is recognisable by its twin-tyred wheels, short barrel and a hydraulic buffer and hydropneumatic recuperator mounted below it, box section split trails and a shield with cut off corners that slopes to the rear. When travelling the barrel is pulled back to the rear out of battery, spades put on top of the trails and a two wheeled limber attached under the rear of the trails.

152mm howitzer M1938 (M-10) in travelling order.

Employment
In service with Romania. Held in reserve by other members of the Warsaw Pact.

130mm M-46 Field Gun

USSR

Calibre: 130mm
Weight: 7,700kg (firing)
8,450kg (travelling)
Length: 11.73m (travelling)
Barrel length: 7.6m (inc muzzle brake)
Width: 2.45m (travelling)
Height: 2.55m (travelling)
Track: 2.06m
G/clearance: 0.4m
Elevation: $-2\frac{1}{2}°$ to $+45°$
Traverse: 50° (total)
Range: 27,150m (HE round max)
1,170m (APHE round max)
Rate of fire: 5-6rpm
Ammunition: HE, projectile weight 33.4kg, m/v 930m/sec
APHE, projectile weight 33.6kg, m/v 930m/sec
Armour penetration: 250mm at 1,000m (APHE round)
Crew: 9
Towing vehicle: AT-S medium tracked artillery tractor

The M-46 was developed from the M1936 Naval Gun and has replaced the M1931/37 (A-19) gun in the Soviet Army. The weapon has a horizontal sliding wedge type breechblock and fires case-type, variable charge, separate loading ammunition.

The weapon is recognisable by its long barrel which has a pepperpot muzzle brake, a small shield (has also been seen without a shield), recoil system above and below the barrel, an inverted U-type collar in front of the shield over the barrel, the trails are of the box type. Each trail has a detachable spade towards the rear of the trail. A two-wheeled limber is provided, when travelling the ordnance is pulled back out of battery until the breech is between the trails.

Employment
Angola, Bulgaria, China (Type 59), Cuba, Czechoslovakia, Egypt, Ethiopia, Finland, Germany (GDR), Hungary, India, Iran, Iraq, Israel, Kampuchea, Libya, Mongolia, Nigeria, North Korea, Peru, Poland, Somalia, Syria, USSR, Vietnam, Yemen (South), Yugoslavia, and Zaire.

130mm field gun M-46 in the firing position.

130mm SM-4-1 Coastal Gun — USSR

Calibre: 130mm
Weight: 16,000kg (firing)
19,000kg (travelling)
Length: 12.8m (travelling)
Barrel length: 7.6m (inc muzzle brake)
Width: 2.85m (travelling)
Height: 3.05m (travelling)
G/clearance: 0.3m
Elevation: –5° to +45°
Traverse: 360°
Range: 29,500m (max)
Rate of fire: 5rpm
Ammunition: HE, projectile weight: 33.4kg, m/v 930m/sec
Towing vehicle: AT-T heavy artillery tractor

The SM-4-1 is a postwar development and is similar to the German SK C/28M mobile coastal gun used during World War 2. Ammunition used with the SM-4-1 is not interchangeable with that used in the 130mm M-46 field gun and KS-30 anti-aircraft gun.

The weapon is mounted on a mobile carriage with four sets of double tyred wheels. It is recognisable by its carriage, muzzle brake, curved shield and the travelling lock for the barrel at the rear of the carriage.

Employment
Egypt, USSR.

130mm SM-4-1 Soviet coastal guns in Egypt.

122mm D-74 Field Gun

USSR

Calibre: 121.92mm
Weight: 5,500kg (firing)
Length: 9.875m (travelling)
Barrel length: 6.599m (inc muzzle brake)
Width: 2.35m (travelling)
Height: 2.745m (travelling)
G/clearance: 0.4m
Elevation: –5° to +45°
Traverse: 58° (total)
Range: 24,000m (HE round max)
1,200m (APHE round max)
Rate of fire: 6-7rpm
Ammunition: HE, projectile weight 27.3kg, m/v 885m/sec
APHE, projectile weight 25kg, m/v 885m/sec
Smoke, illuminating
Armour penetration: 185mm at 1,000m (APHE round)
Crew: 10
Towing vehicle: AT-S medium tracked artillery tractor
AT-L tractor
Ural-375 (6×6) truck

The D-74 122mm field gun was first seen in public in 1955 and was developed as the replacement for the M1931/37 (A-19) 122mm corps gun. It was not built in large numbers, as the M-46 130mm field gun developed at the same time has a much longer range. The D-74 has a semi-automatic vertical sliding wedge breechblock and uses case-type, variable charge, separate loading ammunition. It has a circular firing jack which, with caster wheels (one on each trail, and stowed over the trail whilst travelling), enables the weapon to be quickly traversed through 360°. The carriage of the D-74 is also used for the D-20 152mm gun-howitzer.

It is recognisable by its long stepped barrel and its double-baffle muzzle brake, its recoil system is above the barrel and a part of this is visible just forward of the shield. The shield has a vertically sliding centre section and the sides of the shield slope away in an irregular pattern. The trails are of a box section and split for firing, the circular jack is under the barrel.

Employment
Bulgaria, China (Type 60), Cuba, Egypt, Germany (GDR), Hungary, Nigeria, Poland, Romania and Vietnam.

122mm field gun D-74 in travelling order clearly showing caster wheels.

122mm D-30 Howitzer

USSR

Calibre: 121.92mm
Weight: 3,105kg (firing)
Length: 5.4m (travelling)
Barrel length: 4.875m (inc muzzle brake)
Width: 1.95m (travelling)
Height: 1.66m (travelling)
Track: 1.85m
Elevation: –7° to +70°
Traverse: 360°
Range: 15,400m (HE round max)
1,000m (HEAT round max)
21,000m (RAP round max)
Rate of fire: 7-8rpm
Ammunition: HE, projectile weight 21.8kg, m/v 690m/sec
HEAT, projectile weight 14.1kg, m/v 740m/sec, fin stabilised
Armour penetration: 460mm (HEAT round)
Crew: 7
Towing vehicle: AT-P artillery tractor

ZIL-157 (6×6) truck
Ural-375D (6×6) truck

The D-30 122mm howitzer was developed as the replacement for the M1938 (M-30) and was first identified in 1963. The weapon has a semi-automatic vertical sliding wedge breechblock, and fires case-type, variable charge, separate loading ammunition.

The D-30 is towed muzzle first, some have the towing eye lug underneath the muzzle brake and some have it behind; whilst travelling the three trail legs are clamped under the barrel. When in firing position the crew lower the central jack, raising the wheels off the ground, and then move the outer trails through 120°. The weapon then settles on its trails and can therefore be quickly traversed through 360°. This makes the gun very useful in the anti-tank role.

The weapon is very recognisable by its small shield, muzzle brake, large recoil system above the barrel, and its trails. The ordnance of the 122mm M-1974 self-propelled howitzer is based on the D-30.

Employment
Bulgaria, Cuba, Czechoslovakia, Egypt, Finland, Germany (GDR), Hungary, Israel, Poland, Romania, Syria (some mounted in T-34 tank chassis), USSR, Vietnam and Yugoslavia.

122mm D-30 howitzers in the firing position.

122mm M1938 (M-30) Howitzer — USSR

Calibre: 121.92mm
Weight: 2,450kg (firing)
Length: 5.9m (travelling)
Barrel length: 2.8m
Width: 1.975m (travelling)
Height: 1.82m (travelling)
Track: 1.6m
G/clearance: 0.33m
Elevation: –3° to +63.5°
Traverse: 49° (total)
Range: 11,800m (HE round max)
630m (HEAT round max)
Rate of fire: 5-6rpm
Ammunition: HE, projectile weight 21.8kg, m/v 515m/sec
HEAT, projectile weight 13.3kg, also smoke, chemical and illuminating rounds
Armour penetration: 200mm (HEAT round)
Crew: 8
Towing vehicle: ZIL-151 (6×6) truck
BUCEGI (4×4) truck (Romania)
Valmet Terra (4×4) (Finland)
AT-S medium tracked artillery tractor

This was used by the Soviet Army during World War 2 and was built in large numbers, it is still an effective weapon although it is being replaced by the 122mm D-30 howitzer.

The 122mm M1938 (M-30) has a screw type breechblock and uses case-type, variable charge, separate loading ammunition. The hydraulic buffer is located below the barrel and the hydropneumatic recuperator is above the barrel, no muzzle brake is fitted. The weapon has a shield, the centre of which is raised when in the firing position. The carriage has box section trails with folding spades. It can be fired without spreading the trails, although in this case it has a limited traverse of only 1.5°.

The Bulgarian Army uses M1938s with metal spoked type wheels, the East German Army has some M1938s with much smaller wheels and wider tyres.

Employment
Albania, Algeria, Bulgaria, China (Type 54), Congo, Cuba, Czechoslovakia, Egypt, Finland, Germany (GDR), Hungary, Iraq, North Korea, Poland, Romania, Somalia, Syria, USSR, Vietnam, Yemen (South), and Yugoslavia.

122mm M1938 (M-30) howitzers firing.

122mm M1931/37 (A-19) Corps Gun — USSR

Calibre: 121.92mm
Weight: 7,250kg (firing)
8,050kg (travelling)
Length: 7.87m
Barrel length: 5.645m
Width: 2.46m (travelling)
Height: 2.27m (travelling)
Track: 1.9m
G/clearance: 0.335m
Elevation: –2° to +65°
Traverse: 58° (total)
Range: 20,800m (HE round max)
900m (APHE round max)
Rate of fire: 5-6rpm
Ammunition: HE, projectile weight 25.5kg, m/v 800m/sec
APHE, projectile weight 25kg, m/v 800m/sec
Armour penetration: 190mm at 1,000m (APHE round)
Crew: 8

122mm corps gun M1931/37 (A-19).

Towing vehicle: AT-S medium tracked artillery tractor
AT-T heavy tracked artillery tractor
KrAZ-214 (6×6) truck

This weapon uses the same carriage as the 152mm Gun-Howitzer M1937(ML-20). Wartime models had solid rubber tyres or spoked wheels, now, however, most have twin tyred wheels. It has a hydraulic recoil buffer and hydropneumatic recuperator, a screw type breechblock is fitted. The weapon has been replaced in the Soviet forces by the 122mm D74 and 130mm M-46 Guns.

It is recognisable by its two vertical balancing gears, small shield behind the balancing gears, long barrel with no muzzle brake, it does, however, have a reinforcing band on the muzzle, recoil system is located under the barrel, the barrel is pulled back when in the travelling position. The box section trails are of riveted construction and are provided with detachable picket spades. A two-wheeled limber is provided for travelling purposes.

Employment
Albania, Algeria, Bulgaria, China, Cuba, Czechoslovakia, Egypt, Germany (GDR), Guinea, Hungary, Iraq, Kampuchea, North Korea, North Yemen, Poland, Romania, Somalia, Syria, Tanzania, Vietnam and Yugoslavia.

107mm B-11 Recoilless Gun — USSR

Calibre: 107mm
Weight: 303kg (firing and travelling)
Length: 3.56m (travelling)
Barrel length: 3.383m
Width: 1.46m (travelling)
Height: 0.9m (travelling)
Track: 1.25m
Elevation: −10° to +45°
Traverse: 35°
Range: 6,650m (HE round max horizontal)
450m (HEAT round max effective)
Rate of fire: 5-6rpm
Ammunition: HE, projectile weight 8.5kg, m/v 375m/sec
HEAT, projectile weight 7.5kg, m/v 400m/sec
Armour penetration: 380mm (HEAT round) at 90° angle
Crew: 5
Towing vehicle: ZIL-157(6×6) truck

The 107mm B-11 recoilless gun replaced the B-10 82mm recoilless gun in the Soviet Army, but it has now in turn been replaced by the much lighter 73mm SPG-9 recoilless gun. The B-11 is however still used by many second line and militia units.

The B-11 is a smoothbore recoilless weapon and fires fin stabilised HE and HEAT rounds and has an open type swinging breechblock. In some models the breech has been enlarged and covered with a grill or jacket to protect the crew from the hot chamber.

The wheels can be removed and the gun fired with or without its wheels. It is towed muzzle first or can easily be manhandled.

The B-11 is also known as the RG-107 (RG=Rückstrassfreies Geschütz) and is easily recognisable by its carriage which has the forward leg of the tripod fastened just beneath the tube when in the travelling position and the fixed tubular trails at the rear.

Employment
Bulgaria, China, Egypt, Germany (GDR), Kampuchea, Syria, North Korea and Vietnam.

107mm recoilless gun B-11 in the travelling position.

100mm T-12 Anti-Tank Gun USSR

Calibre: 100mm
Weight: 3,000kg
Length: 9.162m
Barrel length: 8.484m
Width: 1.7m
Height: 1.448m
Elevation: −10° to +20°
Traverse: 27°
Range: 8,500m
Ammunition: APDS, projectile weight 5.5kg, m/v 1,500m/sec
HEAT, projectile weight 9.5kg, m/v 900m/sec
Crew: 6
Towing vehicles: AT-L light tracked artillery tractor
MT-LB multi-purpose tracked vehicle

The 100mm anti-tank gun T-12 has now replaced the older 100mm field gun M1955 in most front line Soviet units. It fires a fin-stabilised non-rotating APDS or HEAT projectile similar in design to those fired by the 115mm gun of the T-62 MBT. In appearance the T-12 is very similar to the 100mm M1955 but the former has a smaller pepperpot muzzle brake that is almost parallel to the muzzle and a different shield. The T-12 can be fitted with an APN-3-5 infra-red night sight. The latest version is designated the MT-12, this weighs 3,100kg.

Employment
The T-12 is known to be used by the GDR, USSR and Yugoslavia and must be considered to be in service with other countries of the Warsaw Pact and the Middle East.

100mm field gun M1944 (BS-3).

100mm field gun M1955.

100mm anti-tank gun T-12.

100mm M1944 (BS-3) Field Gun — USSR

Calibre: 100mm
Weight: 3,650kg (firing)
Length: 9.37m (travelling)
Barrel length: 6.069m (inc muzzle brake)
Width: 2.15m (travelling)
Height: 1.5m (travelling)
Track: 1.8m
G/clearance: 0.33m
Elevation: −5° to +45°
Traverse: 58° (total)
Range: 21,000m (HE round max)
Rate of fire: 8-10rpm
Ammunition: HE, projectile weight 15.7kg, m/v 900m/sec
APHE, projectile weight 15.9kg, m/v 1,000m/sec
HEAT, projectile weight 12.2kg, m/v 900m/sec
Armour penetration: 185mm at 1,000m (APHE round)
380mm at any range (HEAT round)
Crew: 6
Towing vehicle: AT-P artillery tractor
Ural-375D (6×6) truck

This weapon entered service with the Soviet Army towards the end of World War 2 and was developed from a naval gun. In the Soviet Army the M1944 (BS-3) was replaced by the M1955 100mm field gun. A modified version of this gun is used in the T-54 tank (D-10T) and SU-100 assault gun (D-10S). The M1944 has a semi-automatic vertical sliding wedge type breechblock.

It is recognisable by its double-baffle muzzle brake, long barrel, recoil system below the barrel, dual wheels (the wheels have five triangular holes in them), the shield slopes to the rear, the sides of the shield are angled to the side and overhang the wheels. On the front of each side of the shield is a stowage bin.

Employment
Afghanistan, Bulgaria, China (known as Type 59), Congo, Egypt, India, Mongolia, North Korea, Poland, Romania, Somalia, Sudan, Syria (including some mounted on T-34 tank chassis) and Vietnam.

100mm M1955 Field Gun — USSR

Calibre: 100mm
Weight: 3,000kg (firing)
Length: 8.717m (travelling)
Barrel length: 6.126m (inc muzzle brake)
Width: 1.585m (travelling)
Height: 1.89m (travelling)
Track: 1.2m
G/clearance: 0.35m
Elevation: −10° to +20°
Traverse: 27° (total)
Range: 15,400m (HE round max)
Rate of fire: 7rpm
Ammunition: HE, projectile weight 15.7kg, m/v 900m/sec
APHE, projectile weight 15.9kg m/v 1,000m/sec
HEAT, projectile weight 1.22kg, m/v 900m/sec
Armour penetration: 185mm at 1,000m (APHE round)
380mm (HEAT round)
Crew: 6
Towing vehicle: AT-P artillery tractor
ZIL-157 (6×6) truck
Ural-375D (6×6) truck

The 100mm M1955 field gun was the replacement for the 100mm M1944 (BS-3) field gun but it in turn has been replaced by the 100mm anti-tank gun T-12. The latter has a shorter range than the M1955 but better armour penetration.

The M1955 is recognisable by its single road wheels and long stepped barrel which has a pepperpot muzzle brake. The recoil cylinders are on top of the barrel behind the shield. The shield has a flat front with sloping side wings. The split trails are of a box section and a single caster wheel is carried folded on top of the trails whilst travelling. The M1955 can also be fitted with an infra-red sight for night firing.

Employment
Afghanistan, Bulgaria, China, Congo, Czechoslovakia, Egypt, Germany (GDR), Hungary, India, Iraq, Mongolia, Mozambique, North Korea, Poland, Somalia, Sudan, Vietnam and Yugoslavia.

85mm SD-44 Auxiliary-Propelled Field Gun — USSR

Calibre: 85mm
Weight: 2,250kg (firing)
Length: 8.22m (travelling)
Barrel length: 4.693m (inc muzzle brake)
Width: 1.78m (travelling)
Height: 1.42m (travelling)
Track: 1.434m
G/clearance: 0.35m
Elevation: −7° to +35°
Traverse: 54° (total)
Range: 15,650m (max)
Rate of fire: 10-15rpm
Ammunition: HE, projectile weight 9.5kg, m/v 792m/sec
APHE, projectile weight 9.3kg, m/v 792m/sec
HVAP, projectile weight 5.0kg, m/v 1,030m/sec
Armour penetration: 102mm at 1,000m (APHE round)
130mm at 1,000m (HVAP round)
Crew: 7
Towing vehicles: ZIL-157 (6×6) truck
BTR-152 APC
AT-P artillery tractor

The SD-44 is essentially the D-44 Divisional gun fitted with an auxiliary power unit and is recognisable by the engine, steering column and driver's seat which are mounted on the left trail and the ammunition container which is mounted on the right trail. The trails are hollow and carry fuel for the M72 two-cylinder petrol engine, which develops 14hp and gives the SD-44 a maximum road speed of 25km/h and a cross country speed of 8-10km/h. Mounted on either side of the shield, which has an irregular top, is a folding ramrod.

When in the firing position the steering wheel and the trail wheel are folded to enable the left trail spade to dig into the ground, when being towed the third wheel, which is used for steering, is slung between the trail legs.

Employment
Albania, Bulgaria, Cuba, Czechoslovakia, Germany (GDR), Hungary, Poland and Romania.

85mm auxiliary-propelled field gun SD-44 in travelling order.

85mm D-44 Divisional Gun — USSR

Calibre: 85mm
Weight: 1,725kg (firing)
Length: 8.34m (travelling)
Barrel length: 4.693m
Width: 1.78m (travelling)
Height: 1.42m (travelling)
Track: 1.434m
G/clearance: 0.35m
Elevation: –7° to +35°
Traverse: 54° (total)
Range: 15,650m (HE round max)
Rate of fire: 15rpm
Ammunition: HE, projectile weight 9.5kg, m/v 792m/sec
APHE, projectile weight 9.3kg, m/v 792m/sec
HVAP, projectile weight 5.0kg, m/v 1,030m/sec
Armour penetration: 102mm at 1,000m (APHE round)
130mm at 1,000m (HVAP round)
Crew: 8
Towing vehicle: BTR-152 (6×6) APC
AT-P artillery tractor
ZIL-157 (6×6) truck

The D-44 85mm divisional gun was developed as the replacement for the M1942 (ZIS-3) 76mm divisional gun, with the barrel being a development of that used in the T-34/85 tank. The D-44 has a semi-automatic vertical sliding wedge breechblock, hydraulic recoil buffer and a hydropneumatic recuperator.

The D-44-N has an infra-red sight mounted on the top of the shield. Another version is the SD-44 self-propelled (auxiliary) which is described separately.

Its recognition features are its long barrel with a double-baffle muzzle brake, tubular split trails, recoil system (hydraulic buffer and hydropneumatic recuperator) at the rear of the shield. The shield has a top with regular curves and its sides slope to the rear.

Employment
Albania, Algeria, Bulgaria, China (Type 56), Cuba, Egypt, Germany (GDR), Guinea-Bissau, Hungary, Iran, Iraq, Laos, Mali, Morocco, North Korea, Mozambique, Poland, Romania, Somalia, Sudan, Syria, USSR and Vietnam.

85mm divisional gun D-44.

82mm B-10 Recoilless Gun

USSR

Calibre: 82mm
Weight: 72.2kg (firing)
85.4kg (travelling)
Length: 1.91m (travelling)
Barrel length: 1.659m (overall)
Width: 0.714m (travelling)
Height: 0.673m (travelling)
Elevation: −20° to +35°
Traverse: 360°
Range: 4,470m (HE round max)
500m (HEAT round max effective)
Rate of fire: 5-6rpm
Ammunition: HE, projectile weight 4.5kg, m/v 320m/sec
HEAT projectile weight 3.6kg, m/v 322m/sec
Armour penetration: 240mm armour at 90° (HEAT round)
Crew: 4

The B-10 82mm recoilless gun was introduced into the Soviet Army in the early postwar period and for many years was the standard anti-tank weapon of the infantry battalion. It has since been replaced in front line units by the 107mm B-11 which in turn has been replaced by the SPG-9. B-10s are still used

Truck-mounted 82mm recoilless guns of the German (GDR) Army.

in second line and militia units by some members of the Warsaw Pact.

Ammunition fired by the B-10 is of the fin-stabilised type and is loaded from the rear. The PBO-2 optical sight, which is used for both direct and indirect laying is mounted on the left side while the firing mechanism is on the right side. In the past, the B-10 has often been mounted and fired from the rear of a BTR-50P tracked APC.

It has a two-wheeled carriage and is towed by its barrel. A bar-type hand grip is attached to the barrel to assist handling and the weapon can be pulled by two or three men. It can be fired with or without its wheels, in the latter case it is supported by its tripod.

Employment
Bulgaria, China (called Type 65 and is lighter than the original Soviet model), Egypt, Germany (GDR), North Korea, Pakistan, Poland, Syria and Vietnam.

76mm M1966 Mountain Gun — USSR

In the 1960s the Soviet Union introduced into service a new mountain gun which has been given the provisional western designation of the M1966. This weapon has a shield, very short barrel without a muzzle brake and rubber tyred road wheels. No further information on the M1966 is available at the present time. The M1966 is the replacement for the 76mm M1938 Mountain Gun, this weighed 771kg and fired an HE projectile weighing 6.35kg to a maximum range of 10,100m.

Employment
USSR.

76.2mm M1942 (ZIS-3) Divisional Gun — USSR

Calibre: 76.2mm
Weight: 1,116kg (firing)
Length: 6.095m (travelling)
Barrel length: 3.455m (inc muzzle brake)
Width: 1.645m (travelling)
Height: 1.375m (travelling)
Track: 1.4m
G/clearance: 0.36m
Elevation: –5° to +37°
Traverse: 54° (total)
Range: 13,290m (HE round max)
Rate of fire: 15rpm
Ammunition: HE, projectile weight 6.2kg, m/v 680m/sec
APHE, projectile weight at 6.5kg, m/v 655m/sec
HVAP, projectile weight at 3.1kg, m/v 965m/sec
HEAT, projectile weight at 4.0kg, m/v 325m/sec
Armour penetration: 69mm at 1,000m (APHE round)
92mm at 1,000m (HVAP round)
120mm at any range (HEAT round)
Crew: 7
Towing vehicle: BTR-152 APC
Truck (4×4 or 6×6)

The M1942(ZIS-3) 76.2mm divisional gun entered service in 1942 and replaced the earlier 76.2mm M1936, M1939 and M1941 weapons of the same calibre. The carriage of the M1942(ZIS-3) is the same as that of the 57mm M1943(ZIS-2), anti-tank gun, the latter can easily be distinguished from the former by its longer barrel which does not have a muzzle brake. The carriage is of the split tubular trail type with a spade being attached to the end of each trail. The breech block is of the vertical sliding wedge type with the recoil system consisting of a hydraulic buffer and a hydro-pneumatic recuperator.

Its recognition features are its double-baffle muzzle brake, square type shield that projects over the wheels, tubular split trails and hydraulic buffer and hydropneumatic recuperator above and below the barrel.

Employment
Albania, Bulgaria, China (Type 54), Cuba, Czechoslovakia, Egypt, Germany (GDR), Ghana, Indonesia, Kampuchea, Morocco, Nigeria, North Korea, North Yemen, Poland, Somalia, Tanzania, Vietnam and Yugoslavia.

76.2mm divisional gun M1942 (Z15-3).

73mm SPG-9 Recoilless Gun

USSR

Calibre: 73mm
Weight: 47.5kg (launcher)
12.00kg (tripod)
Length: 2.11m
Height: 0.8m
Range: 1,300m (max)
800m (effective)
Armour penetration: 425mm at an incidence of 0°
(HEAT round)
Crew: 3

The 73mm SPG-9 recoilless gun has now replaced the older 107mm B-11 recoilless gun in all front line Soviet units as well as many other members of the Warsaw Pact. In the Soviet Army it is issued on the scale of two per motorised rifle battalion (eg a total of six per tank division and 18 per motorised rifle division) and 18 per airborne rifle division.

The 73mm fin-stabilised HEAT projectile is similar to that fired by the 73mm 2A20 gun mounted in the BMP-1 MICV and has an initial velocity of 435m/sec which increases to 700m/sec. The SPG-9 can be carried by two men for short distances and when in action is supported by a tripod.

The 73mm SPG-9 recoilless gun.

An airborne model called the SPG-9D is also in service, this being provided with a light two wheeled carraige.

Employment

Bulgaria, Germany (GDR), Hungary, Poland and the USSR.

57mm Ch-26 Auxiliary-Propelled Anti-Tank Gun

USSR

Calibre: 57mm
Weight: 1,250kg (firing)
Length: 6.112m (travelling)
Barrel length: 4.07m
Width: 1.8m (travelling)
Height: 1.22m (travelling)
Track: 1.4m
Elevation: −5° to +15°
Traverse: 56° (total)
Range: 6,700m (max)
Rate of fire: 12rpm
Armour penetration: 106mm at 500m (APHE round)
140mm at 500m (HVAP round)
Ammunition: HE, projectile weight 2.8kg, m/v 685m/sec
APHE, projectile weight 3.1kg, m/v 970m/sec
HVAP, projectile weight 1.8kg, m/v 1,240m/sec
Crew: 5
Towing vehicle: (When required) BTR-152 (6×6) APC

On the right hand trail is mounted a M72, two-cylinder, 14hp auxiliary petrol engine, this enables the gun to propel itself at speeds of up to 40km/h on roads. Also on the right hand trail is the steering wheel and the driver's seat. The trails are of a box section with spades on each end, the third wheel is also behind the spades. The weapon can also be towed, when being towed the third wheel is swung up above the trails.

The weapon has a semi-automatic vertical sliding wedge type breechblock, infra-red sighting equipment can be fitted if required.

57mm auxiliary-propelled anti-tank gun Ch-26.

This gun can be recognised by the engine on the right hand trail, drive shaft from the engine to the two wheels. Shield has rounded edges and is lower than the M-1943 (ZIS-2), long barrel with a double baffle muzzle brake and a recoil cylinder underneath it. Each trail also has an ammunition box on it.

Employment

Albania, Cuba, Czechoslovakia, Germany (GDR) (some with the engine removed), Hungary, Poland and Romania.

57mm M1943 (ZIS-2) Anti-Tank Gun — USSR

Calibre: 57mm
Weight: 1,150kg (firing)
Length: 6.795m (travelling)
Barrel length: 4.16m
Width: 1.7m (travelling)
Height: 1.37m (travelling)
Track: 1.4m
G/clearance: 0.315m
Elevation: −5° to +25°
Traverse: 56° (total)
Range: 8,400m (HE round max)
Rate of Fire: 20-25rpm
Ammunition: HE, projectile weight 2.8kg, m/v 706 m/sec
APHE, projectile weight 3.1kg, m/v 990m/sec
HVAP, projectile weight 1.8kg, m/v 1,270m/sec
Armour penetration: 106mm at 500m (APHE round)
140mm at 500m (HVAP-T round)
Crew: 7
Towing vehicle: BTR-152 (6×6) APC
GAZ-69, (4×4) truck
BTR-50P (Carried in)

The weapon has a semi-automatic vertical sliding wedge breechblock and fires the same ammunition as the ASU-57 self-propelled anti-tank gun and the 57mm Ch-26 auxiliary-propelled anti-tank gun. An infra-red night sight can also be fitted.

It is recognisable by its recoil system above and below the barrel, tubular split trails, no muzzle brake, straight topped shield (in recent years, however, some of these weapons have been observed with a wavy top to their shields), fixed spades and a limber loop. The carriage is the same as that used for the 76mm M1942 (ZIS-3) divisional gun.

Employment

Albania, Bulgaria, China (Type 55), Congo, Cuba, Czechoslovakia, Germany (GDR), Egypt, Hungary, North Korea, Poland, Romania, Yugoslavia and Zaire.

57mm anti-tank gun M1943 (ZIS-2).

45mm M1942 Anti-Tank Gun — USSR

Calibre: 45mm
Weight: 570kg (firing)
Length: 4.885m (travelling)
Barrel length: 3.087m
Width: 1.634m (travelling)
Height: 1.2m (travelling)
Track: 1.4m
G/clearance: 0.254m
Elevation: −8° to +25°
Traverse: 30° (left and right)
Range: 4,400m (HE round max)
Rate of fire: 25-30rpm
Ammunition: HE, projectile weight 2.1kg, m/v 343m/sec
APHE, projectile weight 1.4kg, m/v 820m/sec
HVAP, projectile weight 0.9kg, m/v 1,070m/sec
Armour penetration: 66mm at 500m (HVAP round)
Crew: 6
Towing vehicle: UAZ-69 (4×4) truck

The M1942 45mm anti-tank gun is basically the earlier M1937 gun fitted with a new and longer barrel and fires a more powerful APHE projectile. The weapon is recognisable by its split tubular trails, each with a spade, hydropneumatic buffer and spring recuperator recoil system under the long barrel which has no muzzle brake, and the shield which has an irregular top and wings, the top of the shield can be folded forwards to reduce the overall height of the weapon.

The weapon has been seen with two types of tyred wheels, one has plain wheels and the other has triangular holes. The gun has a semi-automatic

vertical sliding wedge type breechblock and a hydropneumatic buffer.

M1942 45mm anti-tank gun.

Employment

Albania and North Korea.

240mm M-240 Heavy Mortar USSR

Calibre: 240mm
Weight: 3,610kg (firing)
Length: 6.51m (travelling)
Barrel length: 5.34m
Width: 2.49m (travelling)
Height: 2.21m (travelling)
Elevation: +45° to +65°
Traverse: 18° (total)
Rate of fire: 1rpm
Ammunition: 100kg, m/v 362m/sec (HE round)
Range: 9,700m (max)
1,500m (min)
Crew: 9
Towing vehicle: AT-P artillery tractor
AT-L light tracked artillery tractor
AT-S medium tracked artillery tractor

This mortar is the largest mortar used by the Soviet Army and was first seen during a parade held in Moscow in November 1953. Until its correct designation became known it was called the M-53 by the West. There have been reports in recent years, however, that this weapon is no longer in service with the Soviet Army.

240mm heavy mortar M-240 being loaded.

It is normally towed muzzle first by a tracked vehicle. A two-wheeled trolley is used to assist loading the ammunition. It is breech loaded and fired by a lanyard. To load the weapon the barrel is swung into the horizontal position in a similar way to the 160mm mortar M-160. The 240mm M-240 is reported to have nuclear capability.

It is recognisable by its very large circular baseplate (213cm in diameter), which has six strengthening pieces, elevating handwheel on the left of the barrel, two vertical cylinders either side of the barrel, collar around the barrel in which the trunions are located and the ground support plate and pickets below and astride the barrel.

Employment

Bulgaria, China, Romania and the USSR.

160mm M1943 Heavy Mortar — USSR

Calibre: 160mm
Weight: 1,170kg (firing)
Length: 3.99m (travelling)
Barrel length: 3.230m
Width: 1.77m (travelling)
Height: 1.41m (travelling)
G/clearance: 0.308m
Elevation: +45° to +85°
Traverse: 25°
Rate of fire: 3rpm
Range: 5,150m (HE round max)
Ammunition: HE, projectile weight 40.8kg, m/v 245m/sec
Crew: 7
Towing vehicle: Truck

This was introduced into the Soviet Army in 1943 and at that time was the largest mortar in service with the Soviet Army. The M1943 was replaced in the Soviet Army by the 160mm M160 Heavy Mortar.

The weapon is of the smooth-bore breech-loading type and is towed muzzle first. The barrel being swung horizontal for loading purposes. It is recognisable by its short barrel, large (152mm) diameter baseplate with small carrying handles, elevating and traversing handwheels are on the left hand side of the barrel at 45°, box-shaped cradle below the barrel, the single-tyred wheels have triangular holes in them.

Employment

Albania, China, Czechoslovakia, Egypt, Germany (GDR), North Korea, Syria and Vietnam.

160mm heavy mortar M1943 in travelling position.

160mm M-160 Heavy Mortar — USSR

Calibre: 160mm
Weight: 1,300kg (firing)
1,470kg (travelling)
Length: 4.86m (travelling)
Barrel length: 4.55m (overall)
Width: 2.03m (travelling)
Height: 1.69m (travelling)
Track: 1.79m
G/clearance: 0.36m
Elevation: +50° to +80°
Traverse: 24° (total)
Rate of fire: 2-3rpm
Range: 8,040m (max)
750m (min)
Ammunition: HE, weight 41.5kg, m/v 343m/sec
Crew: 7
Towing vehicle: BTR-152 (6×6) APC
ZIL-157 (6×6) truck

The 160mm Heavy Mortar M-160 was first seen in 1953 and for some time was referred to in the West as the M-53. The M-160 is no longer in front line service with any member of the Warsaw Pact although some may be in service with mountain units. It is built in China as the Type 56.

To load the mortar the breech is released from the baseplate and swung into a horizontal position, after the round has been loaded it is returned to its original position. It is fired by a lanyard which is attached to the trigger.

It is recognisable by its balancing cylinders below the handwheels and the support arms and anchor pickets below and astride the barrel. It is towed by its muzzle on a two-wheeled carriage, the wheels have five holes in them, although different types of wheels have been seen.

Employment
China, Egypt and Syria.

160mm M-160 heavy mortars of the Egyptian army.

120mm M1938 and M1943 Medium Mortars USSR

Calibre: 120mm
Weight: 274kg (firing)
Barrel length: 1.854m (w/o muzzle attachment)
Elevation: +65° to +80°
Traverse: 8° (total)
Range: 5,700m (max)
460m (min)
Rate of fire: 12-15rpm
Ammunition: Projectile weight 15.4kg, m/v 272m/sec
Crew: 6
Towing vehicle: ZIL-157 (6×6) truck

This mortar is used by most members of the Warsaw Pact and can be found both at battalion level, where it has replaced the 82mm Mortar in some cases, or at regimental level.

The mortar is normally carried horizontally on a two-wheeled tubular trailer, with the baseplate at the rear end. The towing vehicle carries the crew and ammunition. It can also be disassembled into three loads for animal transport.

The projectile can be either (HE) of cast or wrought iron, cast iron has a better effect but a slightly shorter range. The weapon can also fire smoke and incendiary rounds. The barrel is smooth and a device can be fitted to prevent double loading.

Austrian Graw 120mm M-60 mortar in the firing position.

The weapon is recognisable by its circular baseplate which is 106cm in diameter with webbed fins underneath, two legs that are joined by a spring-loaded chain across the bottom, the right leg has an anti-cant device attached to it, twin recoil cylinders extend from the traversing gear to near the travelling clamp.

The M1943 is recognisable by the greater length of the recoil cylinders. Both the M1938 and M1943 fire the same round and can also be fired by trigger as well as normal drop firing.

Note: The M1943 has also been built in China under the designation Type 53 Mortar. The Austrians have a mortar based on the M1943 called the 12cm Granatwerfer M-60.

Employment
Austria, Albania, Czechoslovakia, Egypt, Germany (GDR), Iraq, North Korea, Romania, Syria, Yemen (South) and Yugoslavia.

107mm M1938 Mortar — USSR

Calibre: 106.7mm
Weight: 170kg (firing)
340kg (travelling)
Barrel length: 1.67m
Elevation: +45° to +80°
Traverse: 3°
Rate of fire: 15rpm
Range: 6,300m (max light bomb)
5,150m (max heavy bomb)
Ammunition: HE (light), projectile weight 7.9kg, m/v 263m/sec
HE (heavy), projectile weight 9.00kg, m/v 302m/sec
Crew: 5

The 107mm mortar M1938 was introduced before World War 2 for use by mountain units and is basically a reduced size version of the 120mm M38 mortar. The 107mm mortar is carried complete on a two-wheeled trolly but can also be disassembled for transport by pack animal. It has been replaced in front line Soviet units by an improved version known as the M107.

Employment
China, North Korea, USSR (reserve units) and Vietnam.

82mm M1937, M1941 and M1943 Light Mortars — USSR

Data: M1943
Calibre: 82mm
Weight: 56kg (firing)
Barrel length: 1.22m
Elevation: +45° to +85°
Traverse: 6°
Range: 3,040m (max)
100m (min)
Rate of fire: 15-25rpm
Ammunition: HE, projectile weight 3.315kg
HE, projectile weight 3.087kg
Smoke, projectile weight 3.454kg
Also reported is white phosphorous booster shell with a range of 3,800m
Illuminating — range 2,000m
Crew: 4-5

M1937
The M1936 82mm Mortar was a copy of the M1917/31 Stokes-Brandt Mortar and was developed into the M1937. It is a smooth bore, muzzle loaded mortar and can be broken down into three parts for easy transportation, barrel, baseplate and bipod. The bipod is attached to the barrel near the muzzle end and has two recoil cylinders. Baseplate is circular, 0.58m in diameter and is of welded construction. The current Soviet 82mm mortar is called the new M1937. This has a new lightweight bipod and a new lighter baseplate.

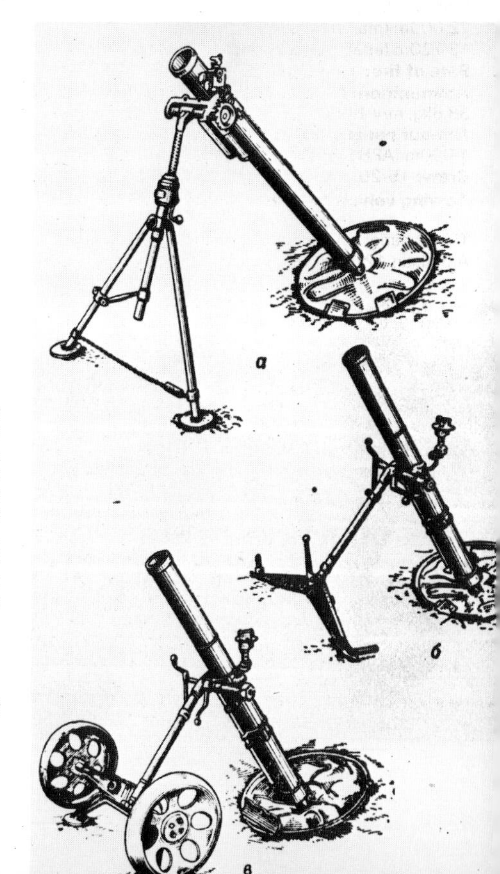

82mm M1937 (top), M1941 (centre) and M1943 (bottom).

M1941
The M1941 can be broken down or towed by its two metal disc type wheels attached to the bipod, these are removed for firing. The bipod is attached midway down the barrel and a recoil cylinder is mounted between the bipod and the baseplate. Weight travelling is 50.3kg, weight firing 45kg, barrel weight 19.5kg, baseplate weight 19kg.

M1943
May be carried or towed. The wheels are not removed for firing, there are support legs on the axle between the two wheels. The recoil cylinder is in a similar position to the M1941 model.

Sights fitted are the MP-1, optical sight and the MP-82 collimeter sight, these also being fitted to the 120mm mortars. The mortars can also be fitted with a muzzle safety device which prevents double loading. It is also reported that these 82mm mortars will fire the 81mm rounds used in British, United States, French and Japanese mortars. The M1937 is also built in Communist China under the designation Type 53, it is also reported that they are, or have been, built in North Korea and the GDR.

The M1941 and M1943 are no longer used by the Soviet Union.

Employment
Albania, Bulgaria, China, Cuba, Czechoslovakia, Congo, Egypt, Germany (GDR), Ghana, Indonesia, Iraq, Kampuchea, North Korea, Syria, Vietnam, North Yemen and Yugoslavia.

130mm KS-30 AA Gun — USSR

Calibre: 130mm
Weight: 24,900kg (firing)
29,500kg (travelling)
Length: 11.521m (travelling)
Barrel length: 8.412m
Width: 3.033m (travelling)
Height: 3.048m (travelling)
Track: 2.388m
G/clearance: 0.408m
Elevation: –5° to +80°
Traverse: 360°
Range: 29,000m (max horizontal)
22,000m (max vertical)
13,720m (effective vertical)
Rate of fire: 10-12rpm
Ammunition: APHE and HE, both projectiles weigh 33.5kg, m/v 1,050m/sec
Armour penetration: 250mm at 0° incidence at 1,000m (APHE round)
Crew: 15-20
Towing vehicle: AT-T heavy tracked artillery tractor

This weapon is similar in many respects to the American 120mm anti-aircraft gun. It was first seen in 1955 but has been replaced by SAM systems in the Soviet Union and Warsaw Pact Forces.

The KS-30 fires fixed charge separate loading ammunition and has a semi-automatic horizontal sliding wedge type breechblock. It is also provided with an automatic fuze setter and an automatic rammer. It is used in conjunction with the PUAZO-30 director and fire control radar SON-30. The gun is mounted on a carriage which has four double tyred wheels, outriggers (folded up when in travelling position) are mounted on the front, rear and either side of the carriage. It has no shield.

Recognition features are its long stepped barrel with a muzzle brake, the barrel is held in place whilst in the travelling position by a lock at the rear of the carriage. The firing platform (at the rear of the gun) folds up at about 45° whilst travelling. Recoil system below the barrel and the long balancing gears under the barrel below the recoil system.

Employment
Vietnam.

130mm AA gun KS-30 in travelling order.

100mm KS-19 AA Gun

USSR

Calibre: 100mm
Weight: 19,450kg (travelling)
Length: 9.238m (travelling)
Barrel length: 5.742m
Width: 2.286m (travelling)
Height: 2.201m (travelling)
Track: 2.165m
G/clearance: 0.305m
Elevation: –3° to +85°
Traverse: 360°
Range: 21,000m (max horizontal)
15,400m (max vertical)
Rate of fire: 15rpm
Ammunition: HE, weight 15.7kg, m/v 900m/sec
APHE, weight 15.9kg, m/v 1,000m/sec
Armour penetration: 185mm at 1,000m (APHE round)
Crew: 9
Towing vehicle: AT-S medium tracked artillery tractor
Ural-375 (6×6) truck

The 100mm anti-aircraft gun KS-19 entered service with the Soviet Army in the late 1940s as the replacement for the 85mm M1939 and M1944 anti-aircraft guns. It is no longer in service in the USSR having been replaced by SAMs.

It has a semi-automatic horizontal sliding wedge type breechblock, an automatic fuze setter and a power rammer. Effective range in the anti-aircraft role is 13,000-14,000m. The gun is usually used with the PUAZO-6/19 director and the SON-9 or SON-9A fire control radar.

The gun is mounted on a four-wheeled carriage with a rectangular firing platform, wheels are removed when in the firing position, folding outriggers are provided. A travel lock is provided on the rear of the carriage.

Recognition features are its long stepped barrel with a multi-baffle muzzle brake, shield in front, twin balancing gears, recoil system above and below the barrel. On the left of the platform is a bracket to carry one round of ammunition for ready use. Above and to the left of the breech is the loading tray, the power gear is on the right side of the gun.

100mm AA gun KS-19 in the firing position.

Employment
Afghanistan, Albania, Algeria, China (known as Type 55), Cuba, Egypt, Germany (GDR), Guinea, Hungary, Iraq, Kampuchea, Morocco, Poland, North Korea, Romania, Somalia, Syria and Vietnam.

85mm M1939 and M1944 AA Guns

USSR

Data: M1939
Calibre: 85mm
Weight: 4,300kg (firing)
Length: 7.049m (travelling)
Barrel length: 4.693m
Width: 2.15m (travelling)
Height: 2.25m (travelling)
Track: 1.80m
G/clearance: 0.4m
Elevation: –3° to +82°
Traverse: 360°
Range: 15,650m (max horizontal)
10,500m (max vertical)
8,380m (effective vertical)
Rate of fire: 15-20rpm
Ammunition: HE, weight 9.5kg, m/v 792m/sec
AP-T, weight 9.3kg, m/v 792m/sec
HVAP weight 5.0kg, m/v 1,030m/sec
Armour penetration: 102mm at 1,000m (AP-T round)
130mm at 1,000m (HVAP round)
Crew: 7
Towing vehicle: ZIL-157 (6×6) truck

The 85mm M1939 anti-aircraft gun, which is also known as the KS-12, entered service with the Soviet Army shortly before the start of World War 2. The ordnance of the M1939 was later used for the T-34/85 tank, SU-85 assault gun and as a towed anti-tank gun. The weapon is no longer in service in the Soviet Union having been replaced by SAMs.

The weapon has a semi-automatic vertical sliding wedge breechblock, a hydropneumatic recuperator and a hydraulic recoil buffer. Sights are provided on the carriage for both anti-aircraft and anti-tank roles.

The M1939 is normally used in conjunction with the PUAZO-6/12 director and the SON/SON-9A (NATO designation 'Fire Can') fire control radar. Recognition features are its long barrel with a conical multi-baffle muzzle brake, it can be seen with or without its shield, recoil system above and below the barrel. The gun is mounted on a pedestal mount with a circular platform that is raised when in the travelling position. It has a four-wheel carriage, a swinging arm either side, each with a jack, and a jack at the front and rear of the carriage to support the carriage when in the firing position.

The 85mm M1939 was succeeded in production by the 85mm M1944 (or KS-18). This has a number of modifications including a longer barrel with a T type muzzle brake rather than the multi-baffle muzzle brake fitted to the original version. The M1944 also fires an HE projectile with m/v of 900m/sec compared to the 792m/sec of the M1939. Maximum effective vertical range of the M1944 is 10,200m compared to the 8,380m of the M1939. In the postwar period production of the M1944 was undertaken in Czechoslovakia.

Employment
Afghanistan, Albania, Algeria, Bulgaria, China (M1939 is known as the Type 56), Cuba, Egypt, Iran, Iraq, North Korea, Poland, Romania, Sudan, Syria, Vietnam, Yemen (South) and Yugoslavia.

85mm AA gun M1939 without shield in travelling order.

57mm S-60 Automatic AA Gun — USSR

Calibre: 57mm
Weight: 4,500kg (firing)
4,660kg (travelling)
Length: 8.5m (travelling)
Barrel length: 4.39m (inc muzzle brake)
Width: 2.054m (travelling)
Height: 2.37m (travelling)
G/clearance: 0.38m
Elevation: –4° to +85°
Traverse: 360°
Range: 12,000m (max horizontal)
8,800m (max vertical)
4,000m (effective on carriage)
6,000m (effective on jacks)
Rate of fire: 70rpm (practical)
Ammunition: HE, projectile weight 2.8kg, m/v 1,000m/sec
APHE, projectile weight 3.1kg, m/v 1,000m/sec
Armour penetration: 106mm at 500m (APHE round)
Crew: 7
Towing vehicle: Ural-375 (6×6)truck
ZIL-157 (6×6) truck

This versatile anti-aircraft gun was first seen in 1950 and is based on the German Type 58 55mm L/70.7 Anti-Aircraft Gun. The gun is the same as that used in the ZSU-57-2 self-propelled AA weapon.

The ammunition is in clips of four and fed into the gun from the left, gun operation is recoil and fully automatic. The ammunition has a proximity fuze fitted.

57mm automatic AA gun S-60 in the firing position.

The S-60 was originally used in conjunction with the PUAZO-5 director and the SON-9/SON-9A ('Fire Can') radar, but the PUAZO-5 has now been replaced by the PUAZO-6/60. More recently an improved integrated system has been introduced for the S-60

The weapon is mounted on a four-wheel carriage that has two folding outriggers and four hand-operated levelling jacks. The wheels are not removed when the weapon is fired.

It is recognisable by its four-wheeled carriage with suspension at 45° to the horizontal, long barrel with a pepperpot muzzle brake, barrel support at the rear of the carriage, front of the shield folds down for travelling and there are three seats to the rear of the weapon.

Employment
Afghanistan, Albania, Algeria, Bulgaria, China (Type 59), Congo, Czechoslovakia, Egypt, Germany (GDR), Guinea, Hungary, Indonesia, Iran, Iraq, Kampuchea, Libya, Mali, Mongolia, Morocco, Mozambique, North Korea, North Yemen, Pakistan, Poland, Romania, Somalia, South Yemen, Syria, USSR, Vietnam and Yugoslavia.

37mm M1939 Automatic AA Gun — USSR

Data: M1939 w/o shield
Calibre: 37mm
Weight: 2,100kg (firing)
Length: 6.036m (travelling)
Barrel length: 2.729m
Width: 1.94m (travelling)
Height: 2.11m (travelling)
Track: 1.545m
G/clearance: 0.36m
Elevation: –5° to +85°
Traverse: 360°
Range: 9,500m (max horizontal)
6,700m (max vertical)
3,000m (effective vertical)
Rate of fire: 80rpm (practical)
Ammunition: HE, projectile weight 0.73kg, m/v 880m/sec
APHE, projectile weight 0.76kg, m/v 880m/sec
HVAP, projectile weight 0.62kg, m/v 960m/sec
Armour penetration: 46mm at 500m (APHE round)
Crew: 8

The M1939 is a modified M1938 light anti-aircraft gun and is based on the Swedish Bofors design and is similar to British and American weapons of the same period. It is no longer in service with the Soviet Army having been replaced by the more powerful S-60 57mm weapon.

The M1939 has a rising breech system with the ammunition being fed from the top in clips of five rounds, underneath the breech is a curved ejection chute for the empty ammunition cases. The hydraulic recoil buffer and the spring recuperator are mounted under the barrel and project a short distance beyond the tube jacket.

The weapon can be seen with or without the shield and is mounted on a four-wheeled carriage, it has been noted that there are two types of disc wheel, one with two holes and the other with five holes. When in the firing position the wheels are removed and the weapon rests on four jacks (one at each end of the carriage, the other two on an outrigger that swings out either side of the carriage), there is also a travelling lock for the barrel at the rear of the carriage.

The Soviets also developed a twin version of the M1939 but this has only appeared outside of the Soviet Union (eg Egypt). The 37mm gun is also used by the Soviet navy in both single (70-K) and liquid cooled twin mounts (V-11M). The Chinese have built the single version as the Type 55 and the twin version as the Type 63.

Employment
Afghanistan, Albania, Algeria, Angola, Bulgaria, China, Congo, Cuba, Egypt, Ethiopia, Germany (GDR), Guinea, Iraq, Mali, Mongolia, Morocco, Mozambique, North Korea, North Yemen, Pakistan, Romania, Somalia, South Yemen, Sudan, Syria, Tanzania, Vietnam, Yugoslavia and Zaire.

Official Soviet drawing of the 37mm automatic AA gun M1939.

23mm ZU-23 Twin Automatic AA Gun USSR

Calibre: 23mm
Weight: 950kg (firing)
Length: 4.57m (travelling)
Barrel length: 2.01m
Width: 1.83m (travelling)
Height: 1.87m (travelling)
Track: 1.67m
Elevation: −10° to +90°
Traverse: 360°
Range: 7,000m (max horizontal)
5,000m (max vertical)
2,500m (effective AA)
Rate of fire: 200rpm/barrel (practical)
Ammunition: HEI-T, projectile weight 0.19kg, m/v 970m/sec
API-T, projectile weight 0.189kg, m/v 970m/sec
Armour penetration: 25mm at 500m (API round)
Crew: 5
Towing vehicle: GAZ-69 (4×4) truck

The ZU-23 has replaced the ZPU-4 in many units of the Warsaw Pact Forces. In the USSR it is only in service with the airborne rifle division. A quad version of this weapon is mounted on a tracked chassis and called the ZSU-23-4, in this case the 23mm guns are water-cooled.

This weapon has quick change barrels and has a cyclic rate of fire of 1,000rd/min/barrel. Operation is gas, fully automatic, with vertical sliding wedge type breechblocks. The twin barrels have flash eliminators fitted. The ammunition is similar to that used in the VYa aircraft cannon, the ammunition boxes are each side of each barrel and each contain 50 rounds of belted ammunition.

Recognition features are its two-wheeled carriage with a small jack at the front and rear and the cylindrical flash eliminators. The wheels go almost flat on the ground when the weapon is in the firing position.

Employment
Afghanistan, Angola, Cuba, Egypt, Ethiopia, Finland, Germany (GDR), Guinea-Bissau, Iran, Iraq, North Korea, Libya, Mozambique, Pakistan, Poland, Somalia, Syria, Tanzania, USSR, Vietnam and South Yemen.

23mm twin automatic AA gun ZU-23 in the firing position.

14.5mm ZPU-4 Quad AA MG USSR

Calibre: 14.5mm
Weight: 1,810kg (firing)
Length: 4.53m (travelling)
Barrel length: 1.348m (w/o flashider)
Width: 1.72m (travelling)
Height: 2.13m (travelling)
G/clearance: 0.458m
Elevation: −10° to +90°
Traverse: 360°
Range: 8,000m (max horizontal)
2,000m (effective horizontal)
5,000m (max vertical)
1,400m (efffective vertical)
Rate of fire: 150rpm/barrel (practical)
600rpm/barrel (cyclic)
Ammunition: API, projectile weight 64.4g m/v 1,000m/sec
Armour penetration: 32mm at 500m (API)

ZPU-4 in the firing position.

Crew: 5
Towing vehicle: GAZ-63 (4×4) truck
GAZ-54 (4×4) truck

This weapon is a quad version of the KPV 14.5mm HMG. Each barrel has 150 rounds of ammunition in a metal link belt in a drum type magazine. The barrels can be quickly changed and system of operation is recoil, full automatic and air cooled.

For ground targets a telescopic sight is provided and for the anti-aircraft role a reflector sight is provided.

The guns are mounted on a four-wheeled carriage with a travelling lock for the guns at the rear, the sight is the highest part of the weapon. When firing the weapon is supported by the four stabilising jacks, screw jack at the font and rear and swinging arms with a jack either side of the carriage. The weapon can be fired with the wheels on the ground.

Recognition features are its four-wheeled carriage, four barrels with drum type ammunition containers and no shield.

Employment
China (Type 56), Egypt, Germany (GDR), North Korea, Somalia, Syria, Tanzania and Vietnam.

14.5mm ZPU-2 Twin Heavy AA MG — USSR

Data: Late production type
Calibre: 14.5mm
Weight: 621kg (firing)
Length: 3.871m (travelling)
Barrel length: 1.348m
Width: 1.372m (travelling)
Height: 1.079m (travelling)
Track: 1.1m
G/clearance: 0.27m
Elevation: −15° to +85°
Traverse: 360°
Range: 8,000m (max horizontal)
5,000m (max vertical)
1,400m (effective vertical)
Rate of fire: 150rpm/barrel (practical)
600rpm/barrel (cyclic)
Ammunition: APT, projectile weight 64.4g, m/v 1,000m/sec
APT-T, projectile weight 59.56g, m/v 1,000m/sec
I-T, projectile weight 59.68g, m/v 1,000m/sec
Armour penetration: 32mm at 500m (APT)
Crew: 4
Towing vehicle: GAZ-69 (4×4) truck

This is a twin version of the Vladimirov (KPV) HMG. There are two models: *Model 1* — the original model with two large mudguards, double tubular towbar, wheels removed for firing and weapon rests on three

points. *Model 2* — later model with two small mudguards, lightweight single towbar and lower silhouette. The wheels are raised when in the firing position.

System of operation is recoil, fully automatic and the ammunition is in metal linked belts of 150 rounds; each barrel has one magazine of 150 rounds.

Its recognition features are its two-wheeled carriage, one large ammunition box on either side, holes for cooling the barrel, flash eliminators and sighting mechanisms high above the barrels. These weapons are also mounted on some K-61 tracked amphibians, BTR-40 and BTR-152 APCs.

Employment
Bulgaria, China (called Type 58), Cuba, Egypt, Germany (GDR), Hungary, North Korea, Poland, Romania, Somalia, Syria, Tanzania and Vietnam.

ZPU-2 14.5mm twin heavy AA gun defending a radar station.

14.5mm ZPU-1 Heavy AA MG — USSR

Calibre: 14.5mm
Weight: 413kg
Length: 3.44m (travelling)
Barrel length: 1.348m
Width: 1.62m (travelling)
Height: 1.34m (travelling)
Elevation: −8° to +88°
Traverse: 360°
Range: 8,000m (max horizontal)
5,000m (max vertical)
1,400m (effective vertical)
Ammunition: APT, projectile weight 64.6g, m/v 1,000m/sec
APT-T, projectile weight 59.56g m/v 1,000m/sec
I-T, projectile weight 59.68g m/v 1,000m/sec
Armour penetration: 32mm at 500m (APT)
Crew: 3
Towing vehicle: GAZ-69 (4×4) truck

This is a single version of the Vladimirov (KPV) heavy machine gun, mounted on a two-wheeled, single-axle carriage. It has quick change barrels and the magazine holds 150 rounds of belted ammunition, this is on the right side. For the AA role it has a reflex optical sight and for the ground role a telescope. It is no longer in service with any of the Warsaw Pact Forces.

Recognition features are its single barrel with flash eliminator, two-wheeled carriage similar to the Model (late) 2 ZPU-2, when in firing position the wheels are raised.

Employment
Vietnam.

5.5in Gun — UK

Calibre: 139.7mm
Weight: 5,850kg (travelling and firing)
Length: 7.518m (travelling)
Barrel length: 4.175m
Width: 2.54m (travelling)
Height: 2.616m (travelling)
Elevation: −5° to +45°
Traverse: 30° left and right
Range: 14,800m with 45.35kg projectile
16,460m with 36.28kg projectile
Rate of fire: 2rpm
Ammunition: HE
Crew: 10
Time into action: 10min
Towing vehicle: 5 ton (6×6) truck

Official designation is Ordnance, BL, 5.5in. The requirement for a 5.5in gun-howitzer was made in 1939. The 5.5in used the same carriage as the obsolete 4.5in gun, the carriage being designed originally for a 4.5in and 5in weapon. The weapon was approved for production in August 1939 and saw action from 1941 in Africa. It was the standard weapon of the medium regiments of the Royal Artillery until the late 1960s and was finally retired from service with the British Army in June 1980.

The weapon is recognisable by its split trails, spades being carried on the top of each trail whilst travelling, large single tyres, barrel has no muzzle brake. The breech mechanism is of the interrupted screw thread type, with the hydropneumatic recoil system being mounted under the barrel, the spring-balancing cylinders are mounted vertically either side of the barrel.

Employment
Australia, Burma, India, New Zealand, Pakistan, Portugal, South Africa and Zimbabwe.

5.5in gun being towed by AEC (6×6) truck.

Recoilless Rifles

UK

During World War II Britain developed a whole range of recoilless weapons. None of these, however, entered service. The first weapon to enter service was the 120mm L1 BAT (Battalion Anti-Tank) in the early 1950s. This was mounted on a two-wheeled carriage and towed by its muzzle. As far as it is known none of these remain in service today.

120mm L4 Mobat
This followed the L1 and was introduced into service in the 1950s and is still used in some numbers by the British Army (reserve units). It is mounted on a two-wheeled carriage and towed by a 4×4 Land Rover by its muzzle. There is an optical sight on the right hand side and a 7.62mm spotting weapon (a modified Bren) on the left hand side. The breech Venturi drops for loading and the round of ammunition is slid into the breech using the top of the venturi. A recent development of the Mobat is the L7 Conbat, this has a 12.7mm ranging MG. Basic data of the Mobat is:
Calibre: 120mm
Weight: 764kg
Length: 4.04m
Width: 1.53m
Height: 1.17m
Elevation: –6° to +30°
Traverse: 360°
Range: 825m
Crew: 3

120mm L6 Wombat
This was developed from the Mobat, is lighter has a longer range, and was introduced into service in the early 1960s. Wombat = weapon of magnesium. The Wombat has replaced the Mobat in many units. It is mounted on a small two-wheeled mount, it is not designed to be towed. It is carried in the rear of a long wheel base Land Rover, and may be fired from the vehicle in an emergency, the vehicle carries six rounds of ready use ammunition. Two small ramps can be placed at the rear of the Land Rover, and with the aid of a winch on the vehicle the Wombat can be quickly unloaded. It can also be mounted and fired from the FV 432 APC. Some of the Volvo Bv202 over-snow vehicles used by the British Royal Marines in Norway have been fitted with an L6 Wombat recoilless rifle.

The Wombat is fitted with an American M8 .50in (12.7mm) spotting rifle over the barrel and the breech opens to the right. The ammunition is the same as that used in the Mobat and the HESH round weighs 27.2kg, has a muzzle velocity of 460m/sec and the projectile itself weighs about 12.8kg. Basic data of the Wombat is:
Calibre: 120mm
Weight: 295kg
Length: 3.86m
Width: 0.86m
Height: 1.09m
Elevation: –8° to +17°
Traverse: 360°
Range: 750m (moving targets)
1,000m (static targets)
Rate of fire: 4rpm
Crew: 3

120mm Mobat of 1st Battalion, the Royal Hampshire Regiment, while the battalion was stationed in Hong Kong.

Employment
L-4 Mobat: Jordan, Kenya, Malaysia, New Zealand and UK.
L-6 Wombat: UK.

120mm Wombat.

120mm Wombat in the ready to fire position on a Land Rover.

105mm Light Gun

UK

Calibre: 105mm
Weight: 1,860kg (travelling and firing)
Length: 7.01m (firing)
4.876m (folded for travelling)
6.324m (travelling, gun forward)
Width: 1.778m
Height: 1.371m (folded for travelling)
2.641m (travelling, gun forward)
Track: 1.397m
Elevation: −5.5° to +70°
Traverse: 5.5° left and right
360° on platform
Range: 17,000m
Rate of fire: 8rpm (for 1min)
6rpm (for 3min)
3rpm (sustained)
Ammunition: HE, projectile weight 16.1kg
HESH, Illuminating, SH/Practice, Smoke (red), Smoke (orange)
Crew: 6
Towing vehicle: 1 tonne (4×4) Land Rover
Bv202 tracked vehicle
Airtransportable: Puma (one load)
Sea King (one load)
Wessex Mk 2 or Mk 5 (two loads)

The 105mm light gun was developed from 1965 by the Royal Armament Research and Development Establishment at Fort Halstead, Kent, as the replacement for the Italian 105mm pack howitzers used by the Light Regiments of the Royal Artillery. After trials, the light gun was accepted by the Army in 1971 and first production guns were delivered in 1974. The light gun has now replaced all 105mm pack howitzers in front line Light Regiments of the Royal Artillery and is also being issued to TA units.

The 105mm gun has two towing attitudes. First is the normal folded position. In this the equipment is jacked up (the jack being stowed on the trail), the quick release, right hand wheel and traverse gear pin removed, and the gun swung through 180°. The barrel is then clamped to the trail and the wheel replaced. This takes less than one minute. To convert from the firing position to the towing attitude the two links are removed, the equipment pushed off the platform, which is then secured to the inside of the trail and the links refitted for travelling. This takes less than one minute. The 105mm gun can be split into two for helicopter loads and reassembled with simple tools in less than 30min. The 105mm light gun can be carried in one load slung under a Sea King or Puma helicopter, the smaller Wessex cannot carry the light gun in one complete lift so the weapon has to be disassembled into two loads.

The barrel has a double baffle muzzle brake and a hand-operated vertical sliding breechblock which is operated by a lever mounted on the top. The hydropneumatic recoil system has a separate recuperator. Direct and indirect sights are provided. The wheels have hydraulic brakes for towing, these can be operated by a lever at the end of the trail whilst firing. The carriage is fitted with trailing arm suspension with shock absorbers. The platform enables the gun to be quickly traversed through 360° and is connected to the underside of the chassis by three wire stays.

A special barrel has been developed to use up the remaining stocks of US M1 ammunition.

Recognition features are its bow-shaped tubular trail with spade, circular traversing platform, no shield, double baffle muzzle brake, balancing springs either side of the rear of the gun, when travelling barrel is to the rear.

Employment
In service with Oman, United Arab Emirates, UK, and seven other countries.

105mm light gun on its turntable.

105mm light gun being hitched up to a one-tonne Land Rover (4×4) vehicle.

25-Pounder Field Gun

UK

Calibre: 87.6mm
Weight: 1,800kg (inc firing platform)
Length: 7.924m (travelling)
Barrel length: 2.35m
Width: 2.12m (travelling)
Height: 1.65m (travelling)
G/clearance: 0.342m
Elevation: −5° to +40°
Traverse: 4° left and right
360° on turntable
Range: 12,250m
Rate of fire: 5rpm
Ammunition: HE, projectile weight 11.34kg, m/v 518m/sec
AP, projectile weight 9.2kg, m/v 609m/sec
Armour penetration: 70mm at 0° incidence at 365m (AP round)
Crew: 6
Towing vehicle: Bedford (4×4) truck

In the late 1920s and early 1930s studies were conducted to find a replacement for the 18-pounder gun and 4.5in howitzer used by the Royal Artillery. The prototype 25-pounder was completed in 1937, this had split trails, but the weapon was found to be too heavy. Instead the carriage of an abandoned 4.1in howitzer was used. The 25-pounder was approved for production in December 1938.

Before sufficient 25-pounders became available 18-pounders were relined and mounted on their original carriages, these were known as 18/25-pounders or 25-pounders Mk 1 and had a range of some 11,700m. The next models were the first complete 25-pounder and the 25-pounder Mk 2, the barrel had no muzzle brake and the trail was of the box type. The next model was the Mk 3, which had a muzzle brake. Over 12,000 25-pounders were built during World War II and the weapons did not pass out of the front line British Army use until 1967. It is, however, still used by the British Army with some Territorial Army units as well as for training.

Employment
Bangladesh, Burma, Cyprus, Ireland, India, Jordan, Kuwait, Pakistan, Portugal, South Africa, Sri Lanka, Sudan, Qatar, United Arab Emirates, UK, South Yemen and Zimbabwe.

25-pounder Mk 3 field gun. Note the turntable under the trail.

17-Pounder Anti-Tank Gun

UK

Calibre: 76.2mm
Weight: 2,923kg (firing)
3,040kg (travelling)
Length: 7.54m (travelling)
Barrel length: 4.442m
Width: 2.225m (travelling)
Height: 1.676m (travelling)
Elevation: −6° to +16.5°
Traverse: 60° (total)
Range: 9,144m (max)
Rate of fire: 20rpm (max)
10rpm (practical)
Crew: 6

The 17-pounder anti-tank gun was developed as the successor to the 57mm 6-pounder anti-tank gun and entered production in 1943, it remained in service with the British Army into the 1950s when it was finally replaced by the 120mm Mobat recoilless anti-tank weapon.

The carriage of the 17-pounder is of the split trail type and the barrel has a double baffle muzzle brake, hydropneumatic recoil system and a vertical sliding semi-automatic breech mechanism. A wide range of fixed ammunition can be fired including AP, APC, APDS and HE, although not all of these are now used. The APDS projectile weighs 3.458kg, has a m/v of 1,203m/sec and will penetrate 321mm of armour at an incidence of 30° at a range of 914m, effective range of the 17-pounder in the anti-tank role is 1,500m.

Employment
Burma, Pakistan and South Africa.

17-pounder anti-tank gun in the firing position from the rear showing the large spades.

6-Pounder Anti-Tank Gun

UK

Calibre: 57mm
Weight: 1,224kg (travelling)
Length: 4.724m (travelling)
Barrel length: 2.565m
Width: 1.889m (travelling)
Height: 1.28m (travelling)
Elevation: −5° to +15°
Traverse: 90° (total)
Range: 8,990m (max)
Rate of fire: 15rpm
Towing vehicle: Jeep

The 6-pounder anti-tank gun was developed shortly before World War II but did not enter production until 1941. It was first used in North Africa in 1941 and was subsequently replaced in the British Army by the much heavier and more powerful 17-pounder anti-tank gun. The gun was also manufactured in the United States as the 57mm gun M1, this was

6-pounder anti-tank gun.

followed by the M1A1 (as M1 but with divided-rim wheels and combat tyres), M1A2 (as M1A1 but with free traverse) and finally the M1A3 (as M1A2 but with modified trail lock and towing eye. The M2 followed the latter in production and incorporated all the previous modifications, but few of these were built.

The carriage of the 6-pounder is of the split trail type with the recoil system being of the hydropneumatic type and the breech mechanism of the vertical sliding wedge type. A wide range of fixed ammunition could be fired including AP, APC, APCBC, APCR, APDS and HE, although not all of these are now in service. The APDS projectile weighed 1.47kg, had a m/v of 1,234m/sec and would penetrate 146mm of armour at a range of 914m. Effective anti-tank range was 1,000m.

Employment
Bangladesh, Burma, Cameroon, Congo, Haiti, India, Pakistan and Rwanda (this includes American supplied M1 series 57mm anti-tank guns).

3.7in AA Gun

UK

Data: Weapon on Mk 3 carriage
Calibre: 93.9mm
Weight: 7,620kg (firing)
9,321kg (travelling)
Length: 8.687m (travelling)
Barrel length: 4.7m
Width: 2.438m (travelling)
Height: 2.510m (travelling)
G/clearance: 0.304m (travelling)
Elevation: −5° to +80°
Traverse: 360°
Range: 18,840m (max horizontal)
12,000m (max vertical)
9,000m (effective vertical)
Rate of fire: 20rpm
Crew: 10-14
Towing vehicle: 5-ton truck

The 3.7in anti-aircraft gun was designed by Vickers-Armstrong shortly before World War II and entered production in 1938, it was also manufactured in Canada. It remained in service with the British Army until the 1950s when it was finally replaced by the Thunderbird SAM which has now been withdrawn from service. There were four types of carriage, Mk 1 (mobile), Mk 2 (static), Mk 3 (mobile) and Mk 4 (mobile).

The weapon has a hydropneumatic recoil system and a horizontal sliding semi-automatic breech mechanism. Two types of ammunition can be fired, AP and HE, both being of the fixed type. The former projectile weighs 12.7kg, has a muzzle velocity of 792m/sec and will penetrate 117mm of armour at an incidence of 0° at a range of 914m. The HE projectile also weighs 12.7kg and has an m/v of 792m/sec.

Employment
Burma, Cyprus, India, Pakistan, South Africa and Yugoslavia.

3.7in AA gun on Mk 3 carriage in travelling order.

40mm Mk 1 Automatic AA Gun

UK

Calibre: 40mm
Weight: 2,034kg (firing)
2,288kg (travelling)
Length: 6.248m (travelling)
Barrel length: 2.249m
Width: 1.92m (travelling)
Height: 2.438m (travelling)
Elevation: −10° to +90°
Traverse: 360°
Range: 4,750m (max horizontal)
4,600m (max vertical)
2,560m (effective vertical)
Rate of fire: 120rpm (cyclic)
60rpm (practical)
Crew: 4-6
Towing vehicle: 6×6 truck

Before World War II the British Army placed an order with the Bofors company of Sweden for a quantity of 40mm m/36 anti-aircraft guns, shortly afterwards a licence was obtained to manufacture the weapon in Great Britain and a further batch of weapons was ordered from Poland who were by then making the Swedish m/36 both for themselves and for the export market. The original Mk 1 carriage was replaced in production by the Mk 2 which was both cheaper and quicker to build.

When in the firing position the carriage is supported on four jacks, one at either end of the carriage and one either side on outriggers. Ammunition is fed to the gun in four-round clips with the empty cartridge cases being ejected via a chute on the forward part of the mount. The shield slopes to the rear and when travelling the barrel is held in position by a travelling lock at the rear of the carriage. The ammunition is of the fixed type with the AP projectile weighing 0.89kg, m/v of 853m/sec, this will penetrate 52mm of armour at a range of 457m or 42mm at a range of 914m, the HE projectile weighs 0.9kg and also has a m/v of 853m/sec.

Employment
Known users include Cyprus, India, Pakistan and Yugoslavia.

3in Mortar

UK

Calibre: 81.84mm
Weight: 18.6kg (barrel, Mk 5)
21.8kg (bipod, Mk 5)
16.3kg (baseplate, Mk 6)
Elevation: +45° to +80°
Range: 450-2,560m
Rate of fire: 15rpm (rapid)
7-10rpm (normal)
Ammunition: HE, projectile weight 4.51kg
Illuminating and Smoke
Crew: 3

The 3in mortar entered service with the British Army in 1936 and was finally replaced by the 81mm mortar in the 1960s. There were five marks of barrel (1, 1A, 2, 4, 5) and six models of baseplate (Nos 1-6).

Employment
Bangladesh, Burma, Cameroon, Egypt, Ghana, India, Indonesia, Malaysia, Nigeria, Pakistan, Saudi Arabia, Sri Lanka and North Yemen.

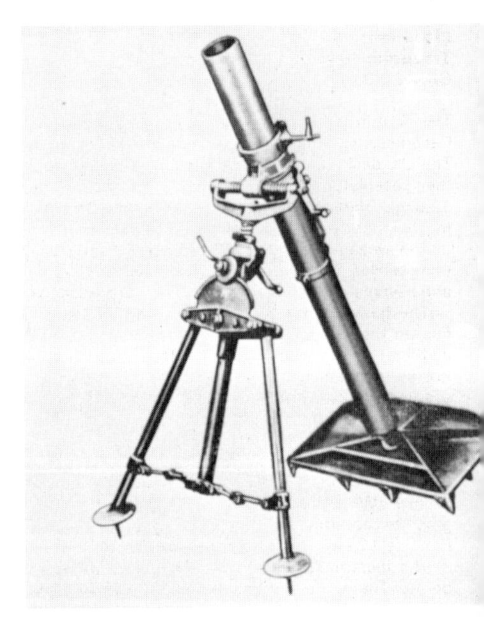

3in mortar in the firing position.

4.2in Mortar

UK

Data: Mk 2
Calibre: 107mm
Weight: 41.5kg (barrel)
36kg (tripod)
550kg (total travelling)
Barrel length: 1.734m
Range: 960-3,750m
Ammunition: Smoke or HE (9.1kg)
Crew: 6

The 4.2in (107mm) mortar was developed in 1940/1941 to fire a chemical round; the first production mortars were completed in 1942. The Mk 1 version had a baseplate and bipod, but the Mk 2 version was fitted with an unusual two-wheeled trailer developed by Jowett Cars Limited. The Mk 2 version was used in the British Army until the 1960s, and was towed by a Land Rover or a 1 ton Truck.

Employment
No longer in service with the British Army but may be employed elsewhere.

British 4.2in Mk 2 mortar in the firing position.

81mm Light Mortar

Canada/UK

Calibre: 81mm
Weight: 11.88kg (barrel)
11.34kg (bipod)
11.36kg (baseplate)
Barrel length: 1.27m (overall)
Bipod length: 1.143m (folded)
Barrel diameter: 86mm (outside muzzle end)
94mm (outside breech end)
Elevation: +45° to +80°
Traverse: 100mils left and right at 45° elevation
Crew: 2-3

The 81mm mortar was designed in the UK and Canada some 16 years ago to replace the 3in mortar. The UK designed the barrel and bipod and Canada the baseplate and sight (Dial Sight C2 weight 1.25kg). The 81mm mortar is more effective and much ligher than the 3in mortar.

The weapon can be broken down into three parts (man packs) for easy transportation, ie barrel, bipod and baseplate.

The barrel is of high-strength steel with cooling fins on the lower half, the bipod is of special steel and light alloy construction. The baseplate is forged in light alloy and has four deep webs, it is circular in shape with its centre dished, in the centre is the socket for holding the barrel.

Ammunition available includes:
High Explosive (HE) L15A3, weight 4.26kg, range 200-5,660m. Body is of ductile cast-iron.
Smoke (WP) L19A4, weight 4.26kg, body is of graphite cast-iron.
Practice, L22A1, range 80m.

The increments are clipped on to the tail of the bomb. There are a total of six primary charges, minimum range of 200m, maximum range 4,500m. There are two new charges, 7 and 8, these increase the range to 5,200m and 5,600m respectively. Rate of fire is 15 rounds per minute. The 81mm mortar is also fitted in the British FV432 APC.

Employment
Austria (8cm Granatwerfer 70), Bahrain, Canada, Guyana, India, Kenya, Malawi, Malaysia, Muscat and Oman, New Zealand, Nigeria, Norway, Qatar, United Arab Emirates, UK and Yemen (North).

81mm medium mortar being prepared for firing.

8in M115 Howitzer

USA

Calibre: 203.3mm
Weight: 13,471kg (firing)
14,515kg (travelling)
Length: 10.972m (overall)
Width: 6.857m (firing)
2.844m (travelling)
Height: 2.743m (travelling)
Elevation: −2° to +65°
Traverse: 30° left and right
Range: 16,800m
Rate of fire: 30rd/h (sustained)
Ammunition: HE (M106), projectile weight 92.53kg, m/v 587m/sec; HE(M404), HE(M509), Agent (M426) and Nuclear (M422)
Crew: 14
Emplacement time: 20min (min)
Towing vehicle: M125 10ton (6×6) truck
M6 tracked vehicle

This weapon was developed during World War II and built by the Hughes Tool Company. It was first known as the 8in Howitzer M1 and Carriage M1. The M115 consists of the M2 Cannon, Recoil Mechanism M4/M4A3 on a M1 Carriage, and the M5 Limber. Some M115s have the M2A1 Cannon which is built of stronger steel.

The M115 has a stepped thread/interrupted screw breechblock, a hydropneumatic variable recoil mechanism and a pneumatic equilibrator.

For travelling the trails are locked together and the two-wheel limber attached. For firing the limber is detached and the weapon supported on the firing jack under the carriage and the trails at the rear. Two spades are carried on either side of each trail and these are placed on the end of each trail when in firing position. Some armies tow the howitzer without the limber.

This weapon is also mounted on a tracked chassis, system is then known as the M110. A wartime model was the M43 but this was built in very small numbers and none remain in service today.

Employment

Belgium, Denmark, Greece, India, Iran, Italy, Japan, Jordan, South Korea, Spain, Taiwan, Turkey and the USA.

8in howitzer M115 of the US Army just after firing with barrel recoilling to the rear.

155mm M198 Howitzer

USA

Calibre: 155mm
Weight: 6,920kg (firing)
6,920kg (travelling)
Length: 11.302m (firing)
12.396m (travelling)
Barrel length: 6.096m
Width: 8.534m (firing)
2.794m (travelling)
Height: 1.803m (min firing)
3.023m (travelling)
G/clearance: 0.33m
Elevation: −5° to +72°
Traverse: 45° (total)
Range: 30,000m (RAP)
Rate of fire: 4rpm
Ammunition: Agent, anti-tank (mine), Cannon Launched Guided Projectile, HE (projectile weight 42.91kg), HE (grenade), HE (mine), Illuminating, RAP, Smoke, Tactical CS, VX and GB
Crew: 10

Towing vehicles: 5ton (6×6) truck
M548 tracked cargo carrier

Since 1941 the standard 155mm medium howitzer of the United States Army has been the M114. This has a max range of only 14,600m so in 1968 the United States Army Armament Command commenced the development of a new 155mm towed howitzer under the designation of the XM198. After extensive trials with prototype howitzers the XM198 was standardised as the M198 and production commenced at Rock Island Arsenal in 1978; production is expected to continue until 1983/84. The M198 is used by the Army and Marines and is deployed in 18-gun battalions with each battalion having three six-gun batteries.

The split trail carriage has a two-position rigid suspension system, when in the firing position the suspension is rotated upwards, and the forward part of the carriage is supported on a non-anchored firing platform. The barrel has a large double-baffle muzzle brake, recoil system is of the hydropneumatic type and the breech mechanism of the screw type.

Employment
USA.

One of the prototypes of the 155mm M198 howitzer, designated the XM198, in the firing position.

155mm M114 and M114A1 Howitzers USA

Calibre: 155mm
Weight: 5,760kg (firing)
5,800kg (travelling)
Length: 7.315m (travelling)
Barrel length: 3.778m (muzzle to end of breech)
Width: 2.438m (travelling)
Height: 1.803m (travelling)
Track: 2.07m
Elevation: −2° to +63°
Traverse: 25° right and 24° left
Range: 14,600m (HE round charge 7)
Rate for fire: 2 rounds in first $\frac{1}{2}$min
8rd in 4min
16rd in 10min
Ammunition: HE, projectile weight 42.9kg, max m/v 563.9m/sec
Agent, HE (grenades), Illuminating, Nuclear, Smoke, Tactical CS, VX and GB
Emplacement time: 5min
Crew: 11
Towing vehicle: M4, M5, M6 traction, 5ton (6×6) truck
Airtransportable: Chinook

The 155mm Howitzer M114 was developed at Rock Island Arsenal before World War 2 and entered production in 1941 as the replacement for the 155mm Howitzer M1918, the latter being the French M1917 built under licence in the United States from 1918. Over 6,000 M114s were built and it remains in service with many armies all over the world, its replacement in the United States Army and Marine Corps is the 155mm M198.

The designation M114 was added after World War 2, the M114 consists of the Cannon, 155mm Howitzer M1 or M1A1, Recoil Mechanism M6 series and Carriage 155mm Howitzer M1A1 or M1A2. The differences are that the M1A1 Cannon is built of stronger steel. The carriage M1A1 has a rack and pinion firing jack and the M1A2 has a screw type firing jack.

The breechblock is of the stepped thread/ interrupted screw type and the recoil mechanism is of the hydropneumatic variable type. The trails are of the split type with a small shield on either side of the barrel, the left had shield top folds down when in the firing position. When in the firing position it rests on three points — both trails and the firing jack. The latter being in front of the shield when in the travelling position, spades are attached to the rear of the trails for firing. For travelling, the trails are locked together and attached to the prime mover.

It is recognisable by its barrel without a muzzle brake, small shields, firing jack and spring type equilibrators each side of the barrel.

It is probable that the United States will fit a number of M114/M114A1s with the ordnance of the new 155mm M198 howitzer, these will be issued to

155mm howitzer M114 in the firing position.

reserve units and designated the M114A2. An auxiliary propelled model of the M114 has been developed under the designation of the M123, but this is in service only in very small numbers. Details of the SRC modified M114 are given under Belgium. The German company of Rheinmetall did develop to the prototype stage a new version of the M114 called the FH155(L), this had a new shield, new ordnance and other modifications but has not so far been adopted. Yugoslavia builds a similar weapon called the M65, available details are given under Yugoslavia. A self-propelled model, the M41, was developed, but this is no longer in service.

Employment

Agentina, Austria, Belgium, Brazil, Denmark, Ethiopia, Germany (FGR), Greece, Iran, Israel, Italy, Japan, Jordan, Kampuchea, Korea (South), Kuwait, Laos, Lebanon, Libya, Netherlands, Norway, Pakistan, Peru, Philippines, Portugal, Saudi Arabia, Spain, Taiwan, Thailand, Turkey, Tunisia, USA, Vietnam and Yugoslavia. Also built in South Korea.

155mm M114 towed howitzer in action in South-East Asia.

155mm M59 Gun

USA

Calibre: 155mm
Weight: 12,600kg (firing)
13,880kg (travelling)
Length: 11.024m (travelling)
Width: 2.512m (travelling)
Height: 2.718m (travelling)
Elevation: −2° to +63°
Traverse: 30° (left and right)
Rate of fire: 1rpm
Range: 22,000m
Ammunition: AP (M112), projectile weight 44.9kg, m/v 854m/sec which will penetrate 76mm of armour HE (M101), projectile weight 43.4kg, and smoke (M104)
Crew: 14
Towing vehicle: 10 ton (6×6) truck
M4 Tractor

In the 1920s development commenced of a 155mm gun which was required to have the same carriage as the 8in howitzer then under development, the latter was eventually standardised as the M1 and redesignated the M115 in the late 1940s.

The 155mm gun M1 was standardised in 1938 and was followed by the M1A1 in 1941, the final production model was the M2, the latter was redesignated the M59 in the late 1940s. The M59 is often called the Long Tom because of its long range and accuracy.

The M59 has a split trail carriage and when in the firing position this is supported on two spades (when travelling these are carried on the top of each trail) and a firing jack. For travelling a two-wheeled limber is attached to the closed trails, but some armies have dispensed with this and hook the weapon directly to the towing vehicles via chains. The ordnance has an hydropneumatic recoil system, interrupted screw breechblock but no muzzle brake.

Employment
Austria, Argentina, Greece, Italy, Japan, Jordan, Korea (South), Pakistan, Spain, Taiwan, Turkey and Yugoslavia.

155mm Long Toms deployed in Italy during World War 2.

106mm M40A1 and M40A1C Recoilless Rifle

USA

Calibre: 106mm
Weight: 113kg (rifle only)
130kg (rifle and spotting rifle)
88kg (M79 mount)
53kg (M92 mount)
Length: 3.4m
Barrel length: 2.845m
Elevation: −17° to +65° (on mount M79)
−17° to +50° (on mount M92)
Traverse: 360°
Range: 700-1,000m (effective)
2,745m (HEAT round, m/v 503m/sec, at 6.5° elevation)
6,876m (HEP-T round, m/v 498m/sec, at 39.5° elevation)
Crew: 2

The M40 was developed from the 105mm M27 recoilless rifle under the designation of the T170/T170E1. This was subsequently standardised as the 106mm recoilless rifle M40 although its actual calibre is 105mm. Late production models are the M40A1C, M40A2 and finally the M40A4.

The M40 is a recoilless, air-cooled breech loaded anti-tank weapon and is aimed using the .50in (12.7mm) M8C spotting rifle which is mounted over the top of the barrel.

The rifle consists of barrel group, breech group, vent assembly and firing cable group:
Barrel group — rifle tube, chamber, quick-breakdown sleeve, spotting rifle front and rear mounting bracket.
Breechblock group — firing pin housing assembly,

106mm M40 recoilless rifle manufactured in India.

breechblock hinge and operating lever dog, cocking cam plate, extractor, sear and breechblock operating lever assembly. The breechblock is of the interrupted thread type.
Firing cable group — spotting rifle and 106mm rifle firing cables, rifle firing operating lever, lanyard and rod assembly.
Vent assembly — consisting of ring assembly, recoil compensating ring and retaining screw.
The M40 can be mounted on any of the following mounts:
Mount M79 — For ground use and for mounting on $\frac{1}{4}$ ton trucks M38 and M38A1C. The M79 has an integral elevating and traversing assembly of the wheelbarrow-tripod type.
Mount M92 — For the M274 Mule light vehicle, it can also be dismounted from the vehicle and fired from the tripod M27. The mount M92 (T173) does not have an integral base assembly.

Employment
Australia, Austria (mounted on a low profile carriage of Austrian design), Brazil, Cameroon, Canada, Chile, Denmark, France, Germany (FGR), Greece, India (manufactured under licence), Iran, Israel (manufactured under licence by Israel Military Industries), Italy, Japan (manufactured under licence as the Type 60), Jordan, Kampuchea, Korea (South), Netherlands, New Zealand, Norway, Pakistan, Philippines, Singapore, Spain (manufactured under licence), Taiwan, Thailand, Turkey and Vietnam.

The M825 is basically an M151 (4×4) 0.75ton light vehicle fitted with a 106mm M40 series recoilless rifle.

105mm XM204 Howitzer USA

The 105mm M102 Howitzer was to have been replaced by the 105mm XM204 Howitzer, development of this was completed several years ago at Rock Island Arsenal and the weapon was standardised as the M204 in 1978. So far no production funding has been authorised and it is unlikely that this weapon will now be placed in production.

105mm M102 Light Towed Howitzer USA

Calibre: 105mm
Weight: 1,496kg (travelling and firing)
Length: 5.182m (travelling)
Barrel length: 3.382m
Width: 1.964m
Height: 1.594m (travelling)
G/clearance: 0.33m
Elevation: −5° to +75°
Traverse: 360°
Range: 11,500m (HE M1)
15,100m (HERA M548)
Rate of fire: 10rpm for first 3min
3rpm sustained
Ammunition: Same as M101A1
Emplacement time: 4min
Crew: 8
Towing vehicle: M561 (6×6)
Airportable: Chinook

The M102 105mm howitzer was developed at Rock Island Arsenal from the late 1950s as the replacement for the M101 105mm howitzers used by the airborne and airmobile divisions. The M102 entered service in 1964 and is employed in 18-gun battalions. Main advantages of the M102 over the earlier M101 are its slightly increased range, the fact that it is 726kg lighter and that it can be quickly traversed through a full 360° enabling it to be relayed on to a new target.

The M102 consists of the Cannon, 105mm Howitzer: M137, Recoil Mechanism, 105mm Howitzer: M37 and Carriage, 105mm Howitzer: M31. It has an aluminium box trail and a hydropneumatic variable length recoil, breechblock is vertical sliding wedge, also has spring equilibrators. The weapon can be quickly traversed through 360° as the carriage pivots around the centre of a circular base by means of a roller (Terra-Tire) located at the rear of the trail assembly. The weapon is staked into position, holes being provided in the firing base for this purpose.

Employment
Kampuchea, USA and Vietnam.

105mm M102 in a camouflaged firing position.

105mm M27 and M27A1 Recoilless Rifle — USA

Calibre: 105mm
Weight: 165kg (rifle only)
153kg (mount and adaptor for M75)
123kg (mount and adaptor for M75A1)
Length: 3.41m
Barrel length: 2.718m
Elevation: −13° to +50°
Traverse: 60° (M75)
80° (M75A1)
Range and Ammunition: HE m/v 341m/sec range 8,560m at 43° elevation
WP m/v 341m/sec range 8,450m at 43° elevation
HEAT (M324) m/v 381m/sec range 3,200m at 9° elevation
HEAT (M341) m/v 503m/sec range 2,740m at 6° elevation
HEP m/v 385m/sec range 3,200m at 9° elevation
HEP-T m/v 515m/sec range 6,970m at 40° elevation
Effective range: 1,000m
Crew: 5, including 2 ammunition members

This is a portable, air-cooled, single-loading, recoilless weapon that fires fixed rounds. Its development designation was T19. Primary role of the M27 is anti-tank. Breechblock is of the interrupted thread type.

The rifle consists of a tube, firing cable, fulcrum ring, breechblock operating lever assembly, chamber, trunnion assembly, trunnion ring and trigger block. The mount consists of the equilibrator and elevating mechanism assembly, travelling lock assembly and top and bottom carriage assembly. With the use of an adapter the mount is designed for truck mounting, the adapter being attached to the bottom carriage. It can be mounted on a variety of vehicles including the M38 or M151 Jeeps or the M29 Cargo Carrier.

The difference between the mount M75 and

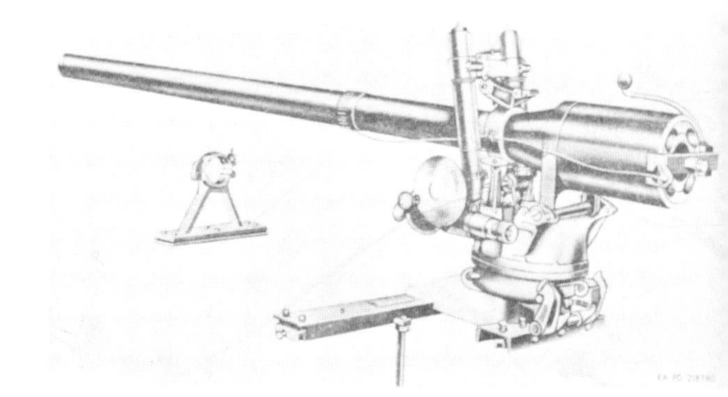

105mm recoilless rifle M27.

M75A1 is in the adapter. The mount M75A1 has an aluminium adapter light enough to be handled by one man and in addition can be attached to three types of jeep. The mount M75 has a steel adapter. A forcing cone was added in the chamber of the rifle M27 making it the M27A1. Telescope fitted is the M94 or M90C. It can also be mounted on the Carriage, Rifle, 105mm M22 (T47), this can be towed by a 4×4 truck and has large tyres.

Employment
France, Israel, Japan, Morocco and Yugoslavia.

105mm M101 and M101A1 Howitzers — USA

Calibre: 105mm
Weight: 2,258kg (firing)
2,258kg (travelling)
Length: 5.991m (travelling)
Barrel length: 2.574m
Width: 3.657m (firing)
2.159m (travelling)
Height: 1.574m (travelling)
Track: 1.778m
Elevation: −5° to +66°
Traverse: 23° left and right
Range: 11,270m
Ammunition: Agent (M60), Agent (M360), APERS-T (XM546), HE(M1), HE (M413), HE(M444), HE(XM710E1), HEP(M327), HERA(M548) (RAP), Illuminating (M314), Leaflet (M84B1), Smoke (M60), Smoke (M84), Tactical CS (XM629), TP-T (M67).
Armour penetration: 102mm at 1,500m (HEAT)
Rate of fire: 8rpm in ½min
4rpm in 4min
3rpm in 10min
100rd/h
Emplacement time: 3min
Crew: 8
Towing vehicle: M5 tractor
M35 or M135 (6×6) truck
Airportable: Chinook

The M101A1 consists of Cannon, 105mm Howitzer, M2A1 Recoil Mechanism M2 and Carriage M2A2. The basic weapon was developed well before the start of World War II, first being known as the gun and carriage M1 (1928), the weapon entered production in 1939 but by 30 June 1940 only 14 had been built, production was completed in 1953 by which time 10,202 had been built.

The weapon has a horizontal sliding wedge breechblock and a hydropneumatic recoil mechanism (constant recoil).

The German firm of Rheinmetall have modified a number of 105mm howitzers for the German Army,

105mm M101 howitzer of the Austrian Army being towed by a 6×6 truck.

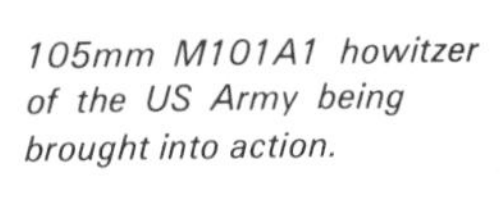

105mm M101A1 howitzer of the US Army being brought into action.

these modifications include a new barrel and new breech ring, new sights and a muzzle brake, range is said to be increased to 14,100m and weight of the weapon is now 2,500kg.

Employment
Argentina, Austria, Australia, Bangladesh, Belgium, Benin, Bolivia, Brazil, Burma, Cameroon, Canada (called C5), Chile, Colombia, Denmark, Dominican Republic, Ecuador, El Salvador, France, Germany (FGR), Greece, Guatemala, Guinea, Haiti, Honduras, Indonesia, Iran, Israel, Italy, Ivory Coast, Japan, Jordan, Kampuchea, Korea (South), Laos, Liberia, Libya, Mexico, Morocco, Netherlands, Nicaragua, Norway, Pakistan, Paraguay, Peru, Philippines, Portugal, Saudi Arabia, Spain, Sudan, Taiwan, Thailand, Turkey, Uruguay, Venezuela, Vietnam and Yugoslavia.

75mm M116 Pack Howitzer USA

Calibre: 75mm
Weight: 653kg (travelling and firing)
Length: 3.658m
Barrel length: 1.195m
Width: 1.194m (travelling)
Height: 0.94m (travelling)
Elevation: −5° to +45°
Traverse: 3° (left and right)
Range: 8,790m (max)
Rate of fire: 22rpm (max)
3rpm (sustained)
Ammunition: HE, projectile weight 8.27kg, max m/v 381m/sec
Smoke and HEAT-T
Crew: 5
Towing vehicle: M37 (4×4) truck or $\frac{1}{4}$ ton truck

Development of this weapon dates from before World War II. The first carriage was the M1, this had box trails and wooden spoked wheels and could be broken down into six loads. This was followed by a carriage with pneumatic tyres, small shield and split trails. The later model was the M8 carriage, this had a box trail and pneumatic tyres. It can be broken down into eight components for mule transport and it takes about 3min to bring into action from this stage. In addition it could also be delivered by parachute. In this case it is broken down into nine components and packed into a paracrate, this takes some 7min to be uncrated, reassembled and emplaced.

The M116 has a horizontal sliding wedge breechblock and a hydropneumatic recoil mechanism. This consists of: Cannon, 75mm Pack Howitzer M1A1, Recoil Mechanism M-1A6, M1A7 or M1A8 on Carriage, Howitzer (Pack) M8. There is also a saluting version of this weapon the M-120, which can only fire blank rounds, welding rod has been used to prevent rounds being placed in the chamber.

After the war the M8 became the M116.

Employment
Bolivia, Brazil, Cameroon, Cuba, Cyprus, Greece, Guatemala, Haiti, Honduras, India, Iran, Japan, Laos, Liberia, Mexico, Morocco, Pakistan, Paraguay, Senegal, Taiwan, Thailand, Turkey, Venezuela, Vietnam and Zaire.

75mm pack howitzer.

75mm M20 Recoilless Rifle

USA

Calibre: 75mm
Weight: 52kg (rifle only)
67kg (rifle and mount M74)
Length: 2.083m (rifle)
Ammunition: projectile weight is 6.53kg
HEAT-T m/v 305m/sec, range 3,200m at 12° elevation
HEP-T m/v 427m/sec, range 6,560m at 42° elevation
HE (TP and WP) m/v 302m/sec, range 6,360m at 43° elevation
Crew: 2-3

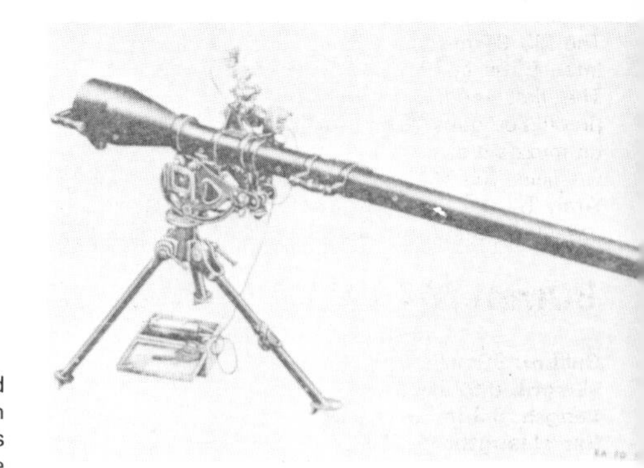

The 75mm M20 recoilless rifle on the M1917A1 mount.

Design of this weapon started in March 1944 and the first pilot was completed and test fired in September 1944, development designation was T25. The first production order for these went to the Miller Printing Machinery Company of Pittsburgh, and production was under way by March 1945.

The 75mm Recoilless Rifle M20 has long been withdrawn from front line service with the United States Army, it still remains in service with many countries around the world, expecially in South America, Africa and the Far East. China has produced the M20 under the designation of the 75mm Recoilless Rifle Type 52.

It has a rifled barrel and is of the air-cooled type and fires fixed rounds, it is capable of direct or indirect fire, sight fitted is the M17A1.

The rifle consists of a barrel group and a breech mechanism group. The breech is of the interrupted thread type. The barrel group consists of a tube, tube handle, mounting bracket, hinge block and vent bushing. The breech mechanism group is attached to the rear end of the barrel group and controls the opening and closing of the breech and the firing of the ammunition. The breech mechanism group consists of the breech operating handle housing, the trigger and firing components of the rifle, the closed breechblock, breechblock hinge and the extractor assembly.

The M20 can be mounted on either the M1917A2 or M74 Mounts, both being of the tripod type. The M1917A2 has a weight of 24kg, elevation −28° to +65° and a traverse of 2° 49min, or 360° when free. The M74 has a weight of 15kg and allows the weapon to traverse through 360°.

37mm M3 Anti-Tank Gun

USA

Calibre: 37mm
Weight: 414kg (travelling and firing)
Length: 3.314m (travelling)
Barrel length: 1.979m
Width: 1.612m (travelling)
Height: 0.958m (travelling)
Elevation: −10° to +50°
Traverse: 60° (total)
Range: 11,750m (max)
500m (effective anti-tank)
Rate of fire: 20rpm (max)
Ammunition: AP, projectile weight 0.87kg, m/v 792m/sec
APC, projectile weight 0.87kg
Canister, weight 0.8kg
HE, projectile weight 0.73kg
Armour penetration: 36mm at 457m (AP round)

37mm anti-tank gun M3 in the firing position.

Crew: 4
Towing vehicle: Jeep (4×4)

The M3 37mm anti-tank gun was developed in the late 1930s and was accepted for service in 1937. The first model had no muzzle brake but later production guns, designated the M3A1, had a screw on muzzle brake although this was often removed in the field. The M3 was replaced in the United States Army by the M1 57mm anti-tank gun which was essentially the British 6-pounder made under licence in the United States. The 37mm gun was installed in many AFVs including the M2 light tank and the M8 Greyhound armoured car.

The M3 has a split trail carriage with a fixed spade on the end of each trail, the recoil system is of the hydro-pneumatic type and the breech mechanism of the vertical sliding wedge type.

Employment
In service with Haiti and possibly Brazil, Chile, Dominican Republic, Mexico and Venezuela.

90mm M117 AA Gun

USA

Calibre: 90mm
Weight: 8,626kg (travelling)
Length: 6.35m (travelling)
Barrel length: 4.728m
(muzzle to rear face of breech ring)
Width: 2.586m (travelling)
Height: 2.845m (travelling)
Track: 2.222m
Elevation: −5° to +80°
Traverse: 360°
Rate of fire: 22rpm
Range: 17,879m (max horizontal)
10,980m (max vertical)
8,500m (effective vertical)
Ammunition: APHE, projectile weight 10.9kg m/v 854m/sec
HE, projectile weight 10.6kg m/v 824m/sec
HVAP, projectile weight 7.63kg, m/v 1,022m/sec
Armour penetration: 147mm at 0° incidence at 1,000m (APHE round)
252mm at 0° incidence at 1,000m (HVAP round)
Towing vehicle: M4 tractor or 7.5ton truck (6×6)

The M117 was previously known as the M1, the M117 designation coming after the last war. The gun and mount were standardised in 1940 and it replaced the 3in M1918 anti-aircraft gun. The M117 consists of the 90mm gun M1A2 or M1A3 (this can be retubed), Mount M1A2 and M1 series Recoil Mechanism.

The breech mechanism is of the vertical sliding type and is semi-automatic, and the recoil mechanism is of the hydro-pneumatic type with a variable recoil mechanism. It has both manual and powered elevating and traversing mechanisms.

The mount consists of a carriage, with a single four-wheeled axle, and outriggers, which fold vertically. Platforms for crew members are carried on the towbar in travelling order. The weapon takes seven minutes to bring into action or 20 minutes if the fire control equipment is to be set up up. The fuze-setter rammer M20 is an integral part of the mount.

Employment
Argentina, Brazil, Greece, Pakistan, Spain, Turkey and Yugoslavia.

Battery of American 90mm M118 anti-aircraft guns on a firing range in West Germany in July 1956.

90mm M118 AA Gun

USA

Calibre: 90mm
Weight: 14,650kg (travelling)
Length: 8.99m (travelling)
Barrel length: 4.496m
Width: 2.62m (travelling)
Height: 3.073m (travelling)
Elevation: –10° to +80°
Traverse: 360°
Rate of fire: 23rpm (sustained)
28rpm (rapid)
Ammunition: APHE, projectile weight 10.9kg, m/v 854m/sec
HE, projectile weight 10.6kg, m/v 824m/sec
HVAP-T, projectile weight 7.63kg, m/v 1,022m/sec
Range: 19,000m (max horizontal)
10,980m (max vertical)
8,500m (effective vertical)

In 1941 the development of a new 90mm anti-aircraft gun to succeed the 90mm M117 commenced, this was required to have a higher rate of fire and be capable of being used in both the ground/ground and coastal defence roles. This weapon was standardised in 1943 as the 90mm Gun M2 and the 90mm Anti-Aircraft Gun Mount M2, in the postwar period the complete weapon was redesignated the M118.

The M118 consists of the 90mm cannon M2A1 or M2A2, recoil mechanism M17 series, mount M2A1 and the fuze-setter rammer M20. The M20 sets the projectile fuzes according to the fire control system or director data and rams rounds into the gun chamber at high speed.

It has a vertical sliding wedge breechblock and a hydropneumatic recoil mechanism. The breech is opened automatically in counter-recoil after firing the first round and closed automatically when the round is rammed. A telescope is provided for use against ground targets.

The weapon is mounted on a four-wheeled carriage. When in the firing position the bogies are removed and the outriggers used. For travelling the front and rear outriggers fold upwards and the side ones go along either side of the carriage. The weapon can however be fired when the bogies and wheels are attached.

Employment
Brazil, Greece, Japan and Turkey.

90mm AA gun M118 in travelling order.

75mm M51 AA Gun

USA

Calibre: 75mm
Weight: 8,750kg (firing)
9,480kg (travelling)
Elevation: –6° to +85°
Traverse: 360°
Range: 13,000m (horizontal)
9,000m (vertical)
Ammunition: HE (proximity fuze), weight 5.7kg, m/v 854m/sec
Rate of fire: 45rpm
Towing vehicle: M8A1 tractor

The M51 75mm anti-aircraft gun was developed from the end of World War 2 and is also known as the Skysweeper. It consists of 75mm Cannon M35, Recoil Mechanism M39, Mount M84 and the M38 Fire Control System.

The M51 can be used both in the anti-aircraft and in the ground role. The M38 fire control system consists of the M15 director, M4 tracking radar (mounted on the left side and having a maximum range of 21,945m), M10 ballistic computer, M16 control power and M22 periscope. Optical sights are provided when the gun is being aimed optically or used in the ground role.

The weapon is mounted on a four-wheeled carriage, the axles and wheels being removed when in the firing position. It is then supported on four retractable outriggers, two of which are vertical when travelling and the other two horizontal either side of the carriage.

The recoil system is hydropneumatic and the recoil length is variable. The breech mechanism is automatic and has a vertical sliding breechblock. The high rate of fire being achieved by having two revolver type magazines at the rear of the weapon, each of which holds six rounds.

Employment
Greece, Japan and Turkey.

75mm AA gun M51 in the firing position.

40mm M1 Light AA Gun

USA

Calibre: 40mm
Weight: 2,656kg (travelling)
Length: 5.728m (travelling)
Barrel length: 2.49m
Width: 1.829m (travelling)
Height: 2.019m (travelling)
Wheelbase: 3.2m
G/clearance: 0.37m
Track: 1.406m
Elevation: –6° to +90°
Traverse: 360°
Range: 4,753m (max horizontal)
4,661m (max vertical)
2,742m (effective vertical)
Ammunition: AP-T (M81), projectile weight 0.89kg, m/v 872m/sec
HE-T, projectile weight 0.935kg, m/v 880m/sec
Rate of fire: 140rpm (cyclic)
70rpm (practical)
Armour penetration: 42mm at 0° incidence at 914m (AP-T round)
Crew: 4-6
Towing vehicle: 4×4 or 6×6 truck

In 1940 the Ordnance Department of the United States Army and the Bureau of Ordnance of the United States Navy both obtained Bofors 40mm light anti-aircraft guns for trials purposes. Trials by the

40mm light AA gun M1 in the firing position.

Army showed that the Bofors gun was superior to the 37mm anti-aircraft gun which had recently been standardised. It became obvious that the 37mm gun could not be produced in anything like sufficient numbers so in 1941 the 40mm anti-aircraft gun was standardised as the M1 (Gun M1 and Carriage M1). The carriage was very complicated and difficult to manufacture so the Firestone Tire and Rubber Company redesigned it and it was later standardised as the M2 carriage. The M2 carriage was followed in production by the M2A1 which had minor improvements. First Bofors guns were delivered to the United States Army early in 1942.

To meet the requirements for a light anti-aircraft gun for air transport, the M5 was developed, but none of these are known to be in service today. At the end of World War 2 the M19 twin 40mm self-propelled anti-aircraft gun was introduced, this being based on the chassis of the M24 light tank. This in turn was replaced in the 1950s by the M42, this has the same turret as the M19 but is based on components of the M41 light tank.

Employment
Known users include Argentina, Brazil, Ecuador, Greece, India, Indonesia, Israel, Italy, Japan, Malaysia, Norway, Pakistan, Portugal, Taiwan, Thailand, Turkey, Vietnam and Yugoslavia.

20mm M167 Vulcan AA Gun — USA

Calibre: 20mm
Weight: 1,588kg (firing and travelling)
Length: 4.04m (firing)
4.7m (travelling)
Barrel length: 1.524m
Width: 2.36m (firing)
1.98m (travelling)
Height: 2.032m (travelling)
Track: 1.764m
Elevation: −5° to +80°
Traverse: 360°
Radar range: 5,000m
Rate of fire: 3,000rpm (high)
1,000rpm (low)
Ammunition: 500 ready rounds of AP-T (M53), HP-T (M54), TP (M55), HEI (M56), TP-T (M220), HEI-T (M242)
Crew: 1 (on weapon)
Towing vehicle: M715, 1.25 ton (4×4) truck
M37, 0.75ton (4×4) truck

The Vulcan Air Defence System was selected by the United States Army after tests in 1964 and 1965, to arm the new Composite Air Defence Battalions. The Vulcan system is in service in two roles: the M167 towed or the M163 on the M113A1 chassis (self-propelled). The Vulcan system was developed by the Armament Systems Department of the General Electric Company of Burlington, Vermont.

The system is made up of the 20mm six-barrelled M168 Vulcan gun, a linked ammunition feed system (on the left hand side of the weapon), and a fire contol system. All mounted in an electrically-powered turret, slewing rates are — azimuth 60° a second and elevation 45° a second. The towed system is provided with an auxilliary power generator mounted on the chassis.

The fire control system consists of a gyro lead-computing gunsight, a range only radar (mounted on the right hand side), and a sight current generator. The gunner visually acquires and tracks the target. The radar supplies range and range rate data to the sight current generator. These inputs are converted to proper current for use in the sight. With this current the sight computes the correct gun lead angle and adds the required super elevation. The system uses the M61 gyro lead-computing gunsight. It is also offered without the ranging radar.

The high rate of fire is used against subsonic aircraft and the lower rate of fire for ground targets such as personnel and light AFVs. The gunner can select 10, 30, 60 or 100-round bursts.

Employment
Belgium, Ecuador, Israel, Jordan, South Korea, Saudi Arabia, USA and Yemen (North).

M167 in the firing position.

12.7mm M55 Quad AA MG

USA

Calibre: 12.7mm
Weight: 975kg (M45C Mount)
363kg (M20 trailer)
Length: 2.89m (travelling)
Width: 2.09m (travelling)
Height: 1.606m (travelling)
1.428m (w/o wheels)
G/clearance: 0.178m
Gradient: 60%
Fording: 0.46m
Elevation: −10° to +90°
Traverse: 360°
Range: 1,500m (effective horizontal)
1,000m (effective vertical)
Rate of fire: 450/550rpm/barrel (cyclic)
Ammunition: Ball (M2 m/v 858m/sec)
AP (M2 m/v 885m/sec)
API (M8 m/v 888m/sec) and others
Crew: 1 (on mount)
Towing vehicle: 2.5ton (6×6) truck

M55 quad .50cal mount in the firing position.

The M55 system consists of the Mount M45C and the Trailer, 1 Ton, Two-Wheel, Machinegun Mount, M20. The mount M45C is power driven and semi-armoured. A power charger produces electrical current to be stored in two 6 volt batteries and the electrical system operates from these batteries. The guns have an azimuth speed of 0-60° a second and in elevation 0-60° a second.

The guns used are the M2 HMG, four of these are used, mounted two each side. Early models had four ammunition chests (two each side), each of the chests held 200 rounds of ammunition, later models had the chests replaced by ammunition box trays. The M20 mount has only two small aircraft type tyres and can only be towed as speeds of up to 16km/h on roads, it is normally carried in the rear of a M35 2½ton (6×6) cargo truck, this vehicle is fitted with equipment enabling the weapon to be deployed off the vehicle. When on the ground the wheels are removed and the weapon is supported on three jacks, one at the drawbar and two at the rear.

An earlier version of this system was the M51, this consisted of the M45 mount and the M17 trailer, the latter had four vehicle type tyres. During World War 2 these weapons and mount M45 were mounted on the US half track, the system was then known as the M16 or M17, depending on the half track used, this vehicle is also still in use. The M45 was developed from the earlier twin .50 MG mount. Further development of the M55 by Israel Aircraft Industries has resulted in the TCM-20 twin 20mm system, full details of which are given under Israel.

Employment
Belgium, Italy, Japan, Jordan, Netherlands, Norway, Pakistan, Portugal, Spain, Thailand, Turkey and Yugoslavia.

107mm M30 (4.2in) Mortar

USA

Calibre: 107mm
Weight: 295kg (complete)
71kg (barrel)
100kg (baseplate M24)
104kg (alternative baseplate)
132kg (mount M24AI w/o baseplate)
146kg (alternative manufacture)
Barrel length: 1.73m
Elevation: +45° to +59°
Traverse: 7° (left and right)
Rate of fire: 5rpm (normal)
20rpm (max)
Ammunition: HE, Illuminating, Smoke (PWP and WP), Chemical (CG, H, HD and HT)
Range:
HE (M329 or M329B1) full charge; 5,420m m/v 293m/sec elevation 45°
HE (M3A1, M3): 4,600m m/v 257m/sec elevation 45°
Chemical (M2 and M2A1): 4,600m m/v 257m/sec elevation 45°
Illuminating (M335): 4,800m m/v 283m/sec elevation 51°
Crew: 5-6

The 107mm mortar M30 was developed by the Chemical Corps under the designation of the T104 and replaced the older 107mm M2 mortar from 1951. The mortar consists of the barrel, standard, bridge assembly, rotator assembly, baseplate and sight unit. The latter was originally the M34A4 but this was subsequently replaced by the M53.

The barrel is closed at the breech end by a tube cap, this has an integral firing pin, the tube cap is supported and aligned on the baseplate and standard assembly, this is equipped with a screw-type mechanism for elevation and traverse.

The mount M24 or M24A1 consists of a rotator assembly, a bridge assembly, a standard assembly and a baseplate. The M24A1 baseplate is one piece steel one and the M24 one is of magnesium with an inner baseplate and an outer baseplate ring.

The M30 is also mounted in the M106 and M106A1 tracked vehicles, these being modifications of the basic M113 and M113A1.

Employment
Austria, Canada, Greece, Iran, Korea (South), Liberia, Netherlands, Norway, Oman, USA and Zaire.

107mm mortar M30 (4.2in) in the firing position.

107mm M2 (4.2in) Mortar

USA

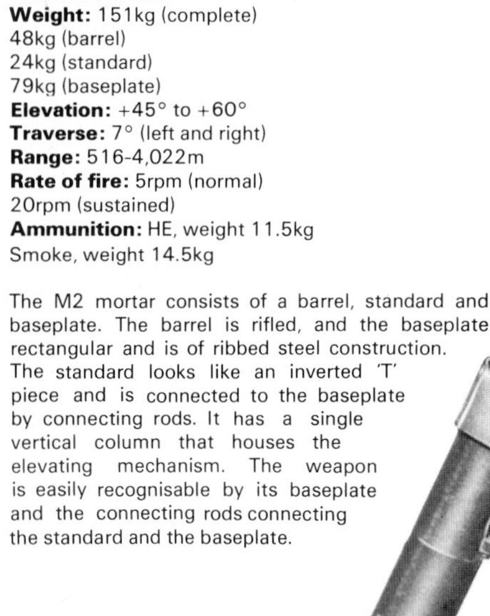

Calibre: 107mm
Weight: 151kg (complete)
48kg (barrel)
24kg (standard)
79kg (baseplate)
Elevation: +45° to +60°
Traverse: 7° (left and right)
Range: 516-4,022m
Rate of fire: 5rpm (normal)
20rpm (sustained)
Ammunition: HE, weight 11.5kg
Smoke, weight 14.5kg

The M2 mortar consists of a barrel, standard and baseplate. The barrel is rifled, and the baseplate rectangular and is of ribbed steel construction. The standard looks like an inverted 'T' piece and is connected to the baseplate by connecting rods. It has a single vertical column that houses the elevating mechanism. The weapon is easily recognisable by its baseplate and the connecting rods connecting the standard and the baseplate.

Employment
Belgium, Ethiopia, India, Italy, Japan, Korea (South), Norway, Pakistan, Philippines, Taiwan and Turkey.

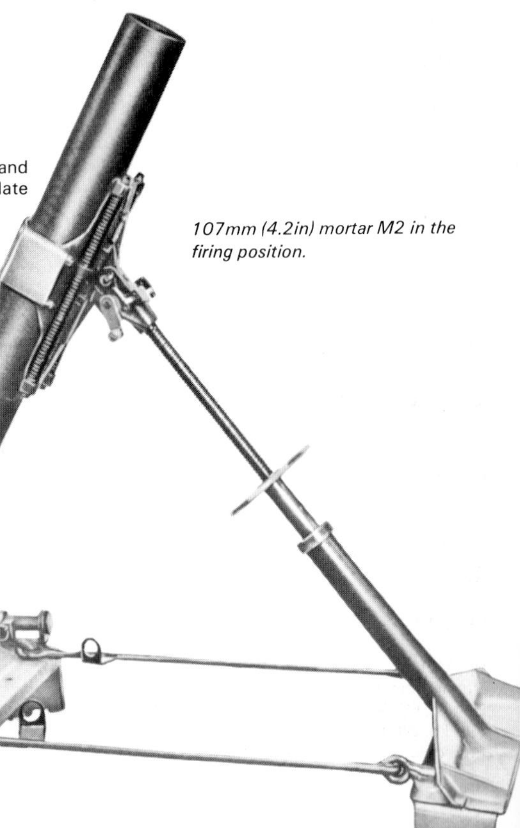

107mm (4.2in) mortar M2 in the firing position.

81mm M29 Mortar

USA

Calibre: 81mm
Weight: 12.70kg (barrel)
21.00kg (baseplate, inner and outer ring)
18.15kg (bipod)
52.20kg (total inc sight)
Barrel length: 1.295m
Elevation: +40° to +85°
Traverse: 4° (left and right)
Rate of fire: 18rpm (normal)
30rpm (max)
Crew: 8 (total)
Ammunition:

Type	*Muzzle Velocity*	*Range*	*Elevation*
HE (M43A1) and TP (M43A1)	211m/sec	3,350m	45°
HE (M56 and M56A1)	174m/sec	2,320m	45°
HE (M362 and M375A2)	234m/sec	3,644m	45°
Illuminating (M301A1 and M301A2)	174m/sec	2,100m	45°
Smoke FS (M57 and M57A1)	174m/sec	2,216m	45°
WP	174m/sec	2,268m	45°

This consists of Cannon: 81mm, M29 and Mount Mortar: 81mm M23 or M23A1. The M29 was developed from the M1 and replaced the M1, the M29 also uses the same ammunition as the M1.

The barrel consists of a tube with a base plug and a fixed firing pin for drop firing and is of the smooth bore type. The exterior of the barrel is helically grooved to reduce weight and this also aids in heat dissipation. The mount consists of a bipod with traversing and elevating mechanisms and a spring type shock absorber above the barrel. There is also a chain between the two legs. The baseplate, which is circular, consists of an inner and outer ring assembly. The M23 has a cross levelling mechanism consisting of a turn buckle and a clamp arrangement. The baseplate has three baseplate latches. The M23A1 baseplate has no latches and its cross levelling mechanism is an adjustable sliding one on the left leg.

Employment
Australia, Austria, Italy, the Netherlands and the USA.

81mm mortar M29.

81mm M1 Mortar

USA

Calibre: 81mm
Weight: 62.0kg (complete)
20.2kg (barrel)
20.4kg (baseplate)
19.27kg (mount)
Barrel length: 1.265m
Elevation: +40° to +85°
Traverse: 5° (left and right)
Rate of fire: 18rpm (normal)
30rpm (max)
Ammunition: HE, Illuminating, Smoke FS and WP
Crew: 2-3 men
Range:

	m/v	*Elevation Angle*	*Range*
HE (M43A1) and TP (M43A1)	211m/sec	45°	3,016m
HE (M56 and M56A1)	174m/sec	45°	2,317m
HE (M362)	234m/sec	45°	2,467m
Illuminating (M301A1 and M301A2)	174m/sec	51°	2,102m
Smoke, FS (M57 and M57A1)	174m/sec	45°	2,216m
Smoke, WP (M57 and M57A1)	174m/sec	45°	2,568m
Projectile, Training, M68	53m/sec	45°	283m

The 81mm mortar consists of Cannon, 81mm, mortar M1 and Mount, Mortar, 81mm, M4. The M1 consists of the tube, base cap and firing pin. The base cap is hollowed and threaded to screw on to the barrel. It ends in a spherical projection, flattened on two sides, this fits into and locks in the socket of the base plate. The firing pin is screwed into the basecap against a shoulder, the firing pin protrudes about 1/20in through the base into the barrel.

The bipod consists of the leg assembly, two chains connect the legs, elevating mechanism assembly and the traversing assembly. The baseplate is rectangular and is of pressed steel, to this has been welded a series of ribs and braces, a front flange, three loops, a link, two carrying handles and the socket.

The weapon can be quickly broken down into three units — barrel, baseplate and bipod for easy transportation. The barrel is smooth bore. The mortar is also mounted in the M4 and M21 half tracks.

Employment
Austria, Belgium, Brazil, Cuba, Denmark, Greece, India, Israel, Italy, Indonesia, Japan, Korea (South), Liberia, Luxembourg, Netherlands, Norway, Pakistan, Philippines, Spain, Taiwan, Thailand, Trinidad, Turkey and Yugoslavia.

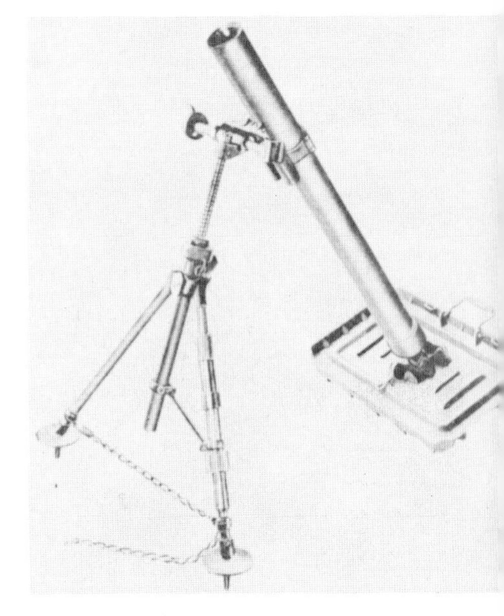

81mm mortar M1.

Cannon-Launched Guided Projectile USA

The 155mm CLGP (commonly known as the Copperhead) is essentially a 155mm HEAT projectile fitted with a laser seeker in the nose and control surfaces. It is handled like a round of conventional ammunition with no modifications being made to the weapon itself.

In 1972 both Texas Instruments and Martin Marietta were awarded development contracts for the CLGP. Each company test fired, from a 155mm M109A1 self-propelled howitzer, a total of 12 CLGPs, Texas Instruments achieved one hit while Martin Marietta achieved eight. Martin Marietta were subsequently awarded a full scale development contract for Copperhead and this is expected to enter service 1982. In addition to being fired from the 155m M109A/A2/A3 self-propelled howitzers it can also be fired from the 155mm towed M198 and NATO weapons such as the 155mm FH-70.

Basically the system works as follows. A laser designator is trained on the target by either a forward observer, from a helicopter or a RPV. The CLGPs semi-active nose mounted laser homes-in with course corrections providing signals to the small control surfaces. These control surfaces provide fly-under, fly-out capabilities to enhance the terminal homing range and manoeuvrability against armour

Sectioned Martin Marietta Cannon-Launched Guided Projectile.

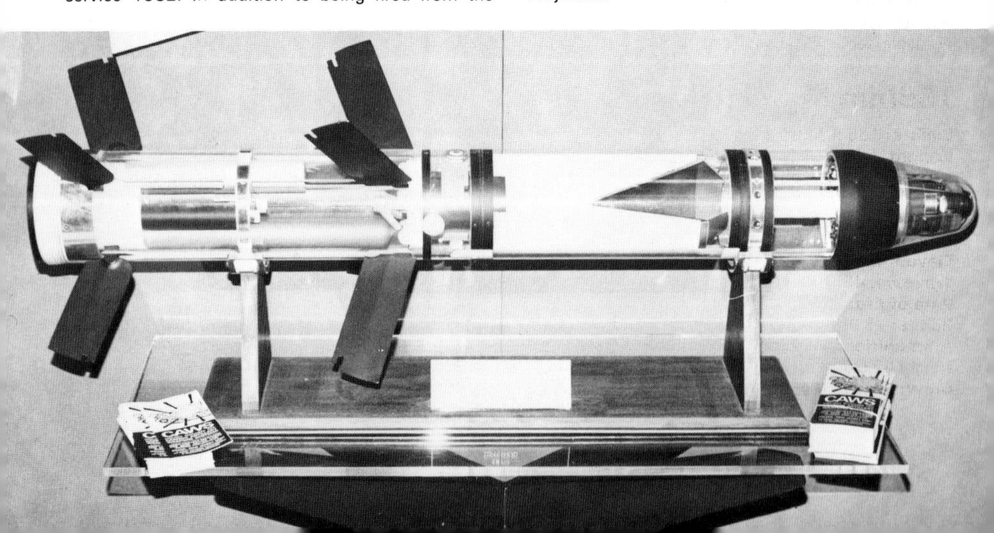

operating under low cloud cover. The CLGP weighs 63.5kg and has a maximum range of 20,000m.

Designators include the Hughes Ground Laser Locator Designator (qv), Laser Target Designator (qv), Modular Universal Laser Equipment (qv), Ferranti Laser Target Marker and Ranger (qv), Lockheed Aquila RPV and the Target Acquisition Designation System mounted in the nose of the Hughes Advanced Attack Helicopter. The Copperhead has been evaluated by the British Army.

Martin Marietta Cannon-Launched Guided Projectile homes on to an M47 tank during trials at White Sands Missile Range.

Employment
In production.

Rocket-Assisted Projectiles

USA

The US Army has had RAPs in service for several years. These have a solid-propellant rocket motor that provides a 2-2.5sec burn so increasing the range. In the case of the 105mm M102, the standard M1 projectile has a max range of 11,500m whereas the M548 RAP (or High Explosive Rocket Assist) has a range of 15,100m. RAPs are also available for 155mm weapons such as the M198 and self-propelled weapons such as the 155mm M109 series and 203mm M110 series. Most countries in the West, and presumably the East as well, are now developing RAPs to increase the range of their artillery. Italy has already developed a 105mm RAP for use with the OTO-Melara 105mm Model 56 Pack Howitzer, and RAPs are being developed for the international FH-70.

155mm M-65 Howitzer

Yugoslavia

Calibre: 155mm *(155mm)*
Weight: 5,500kg *(5,760kg)* (firing)
Length: 8.08m *(7.315m)* (travelling)
Barrel length: 3.797m *(3.778m)*
Width: 2.47m *(2.438m)* (travelling)
Height: 1.8m *(1.803m)* (travelling)
Elevation: −2° to +63°
Traverse: 49° *(49°)* (total)
Rate of fire: 3rpm *(2rpm)*
Range: 15,000m *(14,600m)*
Ammunition: (separate loading) HE, projectile weight 42.9kg m/v 563.9m/sec (M114A1 is same)
Crew: 10 *(11)*

The M-65 155mm howitzer is essentially a modified American M114A1 155mm howitzer made in Yugoslavia. For comparative purposes the data in brackets in the above specification table relate to the original American M114A1 Howitzer.

Employment
In service with Yugoslavia.

105mm M-56 Howitzer

Yugoslavia

Calibre: 105mm
Weight: 2,060kg (firing)
2,100kg (travelling)
Length: 6.17m (travelling)
Barrel length: 3.48m
Width: 2.15m (travelling)
Height: 1.56m (travelling)
G/clearance: 0.335m
Elevation: −12° to +68°
Traverse: 52° (total)
Rate of fire: 16rpm
Range: 13,000m
Ammunition: HE, (semi-fixed), projectile weight 15kg m/v 570m/sec
HEAT (fixed), projectile weight 13.3kg, m/v 395m/sec
Armour penetration: 100mm (HEAT round)
Crew: 6
Towing vehicle: TAM 1500 (4×4) truck

The 105mm Howitzer M-56 is of Yugoslav design and construction and has a number of design features of the American 105mm M101 and the German 105mm M18/40 howitzers, both of which are used in some numbers by the Yugoslav Army.

The M56 has a split trail carriage with a spade being fitted at the end of each trail. The shield is in two parts with each sloping to the side and rear, with an inverted U piece above the ordnance connecting the two halves of the shield.

The ordnance has a multi-baffle muzzle brake, hydraulic recoil buffer, hydropneumatic recuperator and a horizontal sliding wedge breechblock. The M-56 has been observed with two types of road wheels, pressed alloy wheels with solid rubber tyres (similar to those of the German M18) and conventional rubber tyred road wheels (similar to those of the American M101).

Employment
Burma, Indonesia and Yugoslavia.

105mm howitzer M-56 from the rear, this model has American type road wheels.

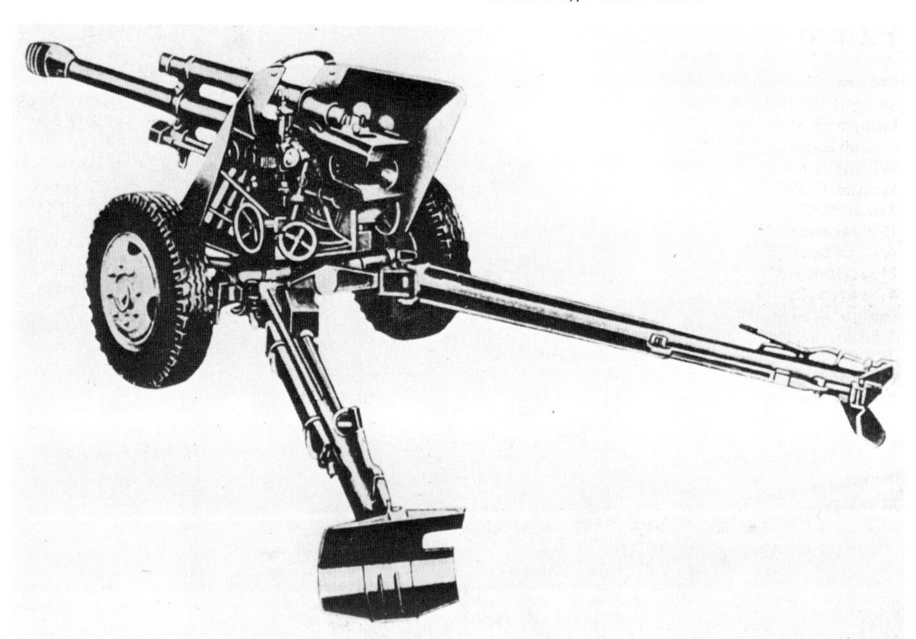

105mm M-65 Recoilless Rifle

Yugoslavia

Calibre: 105mm
Weight: 280kg (firing and travelling)
Length: 4.55m (travelling)
Barrel length: 4.155m
Width: 1.43m (travelling)
Height: 1.14m (travelling)
Elevation: −10° to +30°
Traverse: 360°
Range: 6,000m (max)
1,500m (max effective against stationary targets)
1,000m (max effective against moving targets)
Rate of fire: 4rpm

Crew: 5
Towing vehicle: AR-51 (4×4) light vehicle

The 105mm M-65 recoilless rifle has been developed in Yugoslavia as the replacement for the American supplied 105mm M27 weapon. The M-65 is mounted on a lightweight two-wheeled carriage, some parts of which are believed to be interchangeable with those of the carriage used for the Yugoslav triple 20mm M-55 anti-aircraft gun.

The gunner aims the M-65 with the aid of a 12.7mm UB ranging machine gun, this is used to engage targets out to a range of 600m and is fed by a 20-round belt, the gunner can select either single shots or bursts. The optical sight is used to engage targets out to 1,500m and provision is also made for indirect laying. The M-65 fires a HEAT projectile with a m/v of 455m/sec, this will penetrate 330mm of armour at an incidence of 0°.

Employment
Yugoslavia.

105mm M-65 recoilless rifle showing its shield and the 12.7mm ranging machine gun mounted over the barrel.

82mm M-60 Recoilless Rifle

Yugoslavia

Calibre: 82mm
Weight: 122kg (firing and travelling)
Length: 2.389m (travelling)
Barrell length: 2.2m
Width: 1m (travelling)
Height: 0.827m (travelling)
Track: 0.875m
G/clearance: 0.33m
Axis of bore: 0.5m (firing, low profile)
Elevation: −20° to +35°
Traverse: 360°
Range: 4,500m (max)
1,500m (max effective against stationary targets)
1,000m (max effective against moving targets)
Rate of fire: 4-5rpm
Towing vehicle: AR-51 (4×4) light vehicle

The 82mm M-60 recoilless rifle was developed in Yugoslavia to replace the American supplied 75mm M20 recoilless rifle. The M-60 is mounted on a lightweight two wheeled carriage, the height of which can be varied to aid concealment. In addition the weapon can be disassembled into manpack or animal loads. The PTD M-60 optical sight is mounted on the left side and is graduated up to 1,500m, a mechanical sight is provided for emergency use. The M-60 fires a HEAT projectile weighing 4.3kg, this has a m/v of 388m/sec and will penetrate 220mm of armour at an incidence of 0°.

Employment
In service with Yugoslavia.

82mm M-60 recoilless gun clearly showing the stablising leg and the attachment for fitting the leg to the breech for travelling.

76mm M-48 Mountain Gun

Yugoslavia

Calibre: 76.2mm
Weight: 720kg (travelling)
Length: 2.42m
Barrel length: 1.178m
Width: 1.46m
Height: 1.22m
G/clearance: 0.18m
Elevation: –15° to +45°
Traverse: 50°
Rate of fire: 25rpm
Range: 8,750m
Ammunition: (semi-fixed) HE, projectile weight 6.2kg m/v 398m/sec
HEAT, projectile weight 5.1kg
Armour penetration: 100mm at 0° incidence at 450m (HEAT round)
Crew: 6
Towing vehicle: Zastava AR-51 (4×4) light vehicle

The M-48 76mm mountain gun has been developed to meet the requirements of Yugoslavia mountain units for a weapon which could be quickly disassembled for transport by pack animal as well as being towed by animals or a light vehicle. The M-48 is often called the Tito Gun.

There are at least four models of the M-48:
M-48 (B-1) which can be towed at a maximum road speed of 60km/h or disassembled into eight pack loads, or towed by two animals in tandem.
M-48 (B-1A1-1) has same pneumatic wheels and tyres as the M-48(B-1) but has some suspension parts of the M-48(B-1A2), this model cannot be disassembled or towed by animal.
M-48(B-1A2) has light alloy wheels with solid rubber tyres, maximum towing speed is 30km/h.
M-48(B-2) is the latest model, no details of this are available at the present time.

Recognition features are its very short barrel which has a multi-baffle muzzle brake, shield with wings that slope to the rear and split trails, when travelling the rear half of the latter fold forwards through 180° on to the forward part of the trail.

Employment
Burma, Indonesia (unconfirmed), Sri-Lanka and Yugoslavia.

76mm mountain gun M-48 in the firing position.

20mm M-55 AA Gun

Yugoslavia

Calibre: 20mm
Weight: 1,089kg (firing)
1,171kg (travelling)
Length: 4.04m (travelling)
Barrel length: 1.956m
Width: 1.27m (travelling)
Height: 1.93m (travelling)
Elevation: –5° to +83°
Traverse: 360°
Range: 1,500-2,000m (effective AA)
Rate of fire: 700/800rpm/barrel (cyclic)
150rpm/barrel (practical)
Crew: 6 (1 on gun)
Towing vehicle: AR-51 (4×4) light vehicle

The 20mm M-55 anti-aircraft gun is basically the Swiss HSS630-3 carriage fitted with the HSS804 cannon, all of which is manufactured under licence in Yugoslavia for the home and export markets.

When in the firing position the wheels are raised off the ground and the carriage is supported on three outriggers, one at the front and two at the rear, if required it can be fired with the wheels in the travelling position. Each 20mm cannon has a 50-round drum of ready use ammunition. Types of ammunition that can be fired include AP, API, API-T, HE, HEI, HEI-T and TP. The AP projectile weighs 142g, has an m/v of 840m/sec and will penetrate 18mm of armour at an incidence of 0° at a range of 500m. The HE projectile weighs 122g and has an m/v of 880m/sec.

There are at least two variants of the M-55, the M-55A3B1 with powered traverse and elevation and an optical mechanical sight, and the M-55A4B1 also with powered traverse and elevation but with a computer/hydraulic sight. It is probable that the sights are manufactured under licence from Officine Galileo of Italy.

Employment
Mozambique and Yugoslavia.

The basic 20mm M-55 AA gun in travelling order being towed by an AR-51 (4×4) light vehicle.

120mm UBM-52 Mortar

Yugoslavia

Calibre: 120mm
Weight: 400kg (firing)
420kg (travelling)
Barrel length: 1.29m
Elevation: +45° to +85°
Traverse: 6°
Rate of fire: 15-25rpm
Range: 6,010m with light mortar bomb (12.25kg)
4,760m with heavy mortar bomb (15.9kg)
Ammunition: HE, Smoke and Illuminating
Crew: 5
Towing vehicle: AR-51 Zastava (4×4) truck

This 120mm mortar is of Yugoslav design and construction and easily recognisable by its medium sized road wheels that have solid tyres, these wheels have 14 triangular holes in each of them. Each wheel is provided with two eyes through which ropes can be put for carrying purposes.

When in the travelling position the mortar is towed by its muzzle end with the barrel horizontal, the towing eye folds down on top of the barrel when not in use.

When required for firing the link between the webbed baseplate and the axle of the mount is broken enabling the barrel to go to the rear.

Employment
Burma, Indonesia and Yugoslavia.

120mm mortar UBM-52 in the firing position.

81mm M31 Mortar

Yugoslavia

Calibre: 81mm
Weight: 61.5kg (firing)
21kg (barrel)
19kg (bipod)
20kg (baseplate)
1.5kg (sight)
Barrel length: 1.31m
Range: 85m (min)
4,100m (max)
Rate of fire: 20rpm
Ammunition: HE (light), 3.2kg, HE (heavy), 4.2kg, Smoke and Illuminating

The Yugoslav 81mm Mortar M31 is basically a copy of the American 81mm mortar M1 although the Yugoslav weapon has a longer range. It consists of four main components, rectangular baseplate, barrel, bipod and sight.

Employment
Yugoslavia.

81mm M68 Mortar

Yugoslavia

Calibre: 81mm
Weight: 41.5kg (firing)
16kg (barrel)
13kg (bipod)
11kg (baseplate)
1.5kg (sight)
Range: 5,000m (max)
Rate of fire: 20rpm
Ammunition: HE, weight 3.3kg

The M68 81mm mortar is of Yugoslav design and construction although it is very similar to the French Hotchkiss-Brandt MO-81-61L mortar. The M68 has now replaced the M31 in Yugoslav units and is both lighter and has a longer range.

Employment
Yugoslavia. Probably exported to other countries.

Part Two: Multiple Rocket Systems

Argentinian Multiple Rocket Systems

The Instituto de Investigaciones Cientifas Y Tecnicas de las Fuerzas Armadas (CITEFA) has developed two multiple rocket systems to meet the requirements of the Argentinian Army, these are called the Pampero and the SAPBA-I.

Pampero
This 105mm 16-round multiple rocket system was at the pre-production phase late in 1979. The solid propellant rockets have four folding fins and weigh 30kg at launch with the warhead weighing 10kg. Max range is 12,000m and max dispersion for a 16-round multiple launch is 300m × 200m.

SAPBA-I
This 127mm 40-round multiple rocket system is now in its final development phase. The solid propellant rockets weigh 56kg and have four folded fins. Max dispersion for a 40-round multiple launch is 500m × 400m.

130mm M51 (32-round) MRS on Steyr 680 M3 Chassis

Austria

Weight: 8,400kg (system loaded)
Length: 6.73m (travelling)
Width: 2.4m (travelling)
Height: 2.9m (travelling)
G/clearance: 0.3m
Engine: Steyr WD610.71 6-cylinder supercharged diesel developing 165hp at 2,800rpm
Speed: 79.7km/h
Range: 500km
Fuel: 180 litres
Fording: 0.5m
Gradient: 74%
Elevation of launcher: 0° to +50°
Traverse of launcher: 240° (total)
Calibre of rocket: 130mm
No of launcher tubes: 32
Length of rocket: 0.8m
Weight of rocket: 24.2kg (complete)
Range: 8,200m (max)
Velocity: 410m/sec (max)
Time to reload: 2min
Crew: 6

Some years ago Austria purchased a quantity of M51 multiple rocket systems from Czechoslovakia mounted on the standard Praga V3S (6×6) truck chassis. The Austrians have recently removed these from their original chassis and fitted them to the Steyr 680 M3 (6×6) chassis as this is one of the standard vehicles of the Austrian Army. Reserve rockets are carried in stowage boxes below and either side of the launcher.

Employment
Austria.

130mm M51 (32-round) MRS on a Steyr 680 M3 (6×6) truck chassis.

Brazilian Rocket Systems

Since the 1960s Brazil has undertaken an extensive rocket development programme which has so far resulted in the development of at least three unguided surface-to-surface rockets, one of which has already entered service with the Brazilian Army. The three rockets are designated the 108R, X20 and X40.

108R
This rocket has a single stage solid propellant motor and is spin stabilised by the canted nozzles of the motor. The rockets are launched from the X2 16-barrelled launcher which can be mounted on the rear of a truck or trailer.

X20
This rocket has a single stage solid-propellant motor and is spin stabilised by cruciform tail fins with the rockets being launched from a three-rail launcher.

X40
This rocket has a single stage solid-propellant motor and is spin-stabilised by cruciform tail fins. The X40 can be launched from a single rail launcher or a three rail launcher mounted on the top of a Brazilian X1A tank chassis.

Future Developments
These include the development of a X300 rocket with a range of some 350km as well as the much longer range X100 rocket.

Rocket Designation:	*108R*	*X20*	*X40*
Calibre of rocket:	108mm	180mm	300mm
Length of rocket:	0.93m	2.9m	4.45m
Weight of rocket: (complete)	16.8kg	116kg	550kg
Weight of warhead:	3kg	35kg	146kg
Range: (max)	7,500m	35,000m	68,000m
No of rockets in system:	16	3	1 or 3
Launcher:	trailer/vehicle	trailer	trailer/vehicle

Employment
The 108mm 108R is in service with the Brazilian Army.

X40 rockets ready to be launched, the carrier vehicle is a modified X1A tank chassis.

107mm Type 63 (12-round) Multiple Rocket Launcher

China, People's Republic of

Weight of system: 602kg
Length: 2.895m (travelling)
Width: 1.651m (travelling)
Height: 0.91m (travelling)
Elevation of launcher: −3° to +57°
Traverse of launcher: 32° (total)
Calibre of rocket: 106.7mm
No of launcher tubes: 12
Length of rocket: 0.837m (complete)
Weight of rocket: 19kg (complete)
Range: 8,050m (max)
Velocity: 385m/sec (max)
Time to reload: 3min
Crew: 5

The 107mm (12-round) Type 63 multiple rocket system was developed as the replacement for the older 102mm (6-round) Type 50-5 system which is no longer in service.

The launcher has three banks of four barrels and is mounted on a split pole type carriage that is provided with two pneumatic road wheels. When in the firing position the wheels are removed and the launcher is supported by the two trails at the rear, each of which is fitted with a spade when in the firing position, and two short legs at the front of the carriage.

A pack model of the Type 63 has been developed for use by airborne and mountain units, this weighs 281kg when in the firing position. There is also a lightweight model of the Type 63 designated the Type 63-1, this has spoked road wheels and four banks of three rockets. A single 106.7mm rocket launcher has also been developed for use by guerilla units.

Employment
Albania, China and Vietnam.

107mm (12-round) Type 63 multiple rocket launcher in travelling order.

130mm M51 (32-round) MRS on Praga V3S Chassis

Czechoslovakia

Weight of system: 8,900kg (loaded)
Length: 6.1m (travelling)
Width: 2.31m (travelling)
Height: 2.92m (travelling)
G/clearance: 0.4m
Engine: Tatra T-912 6-cylinder air-cooled diesel developing 98hp at 2,100rpm
Speed: 62km/h
Range: 443km
Fuel: 120 litres
Fording: 0.8m
Gradient: 60%
Elevation of launcher: 0° to +50°
Traverse of launcher: 240° (total)
Calibre of rocket: 130mm
No of launcher tubes: 32
Length of rocket: 0.8m
Weight of rocket: 24.2kg (complete)

Battery of 130mm M51 (32-round) MRSs firing.

Range: 8,200m (max)
Velocity: 410m/sec (max)
Time to reload: 2min
Crew: 6

The 130mm M51 (32-round) multiple rocket system is basically a standard Praga V3S (6×6) truck chassis with a 32-round rocket launcher mounted to the rear of the cab, as the latter is not provided with any form of armour protection, the launcher is traversed left or right prior to the rockets being launched. The launcher is also used by the Austrian Army mounted on the rear of a Steyr 680 M3 chassis (qv) and Romania mounted on a Soviet ZIL-151/ZIL-157 chassis. The M51 is no longer in production and is being replaced in Czechoslovakian units by the 122mm M1972 (40-round) multiple rocket system on Tatra 813 chassis.

Employment
Austria, Bulgaria, Czechoslovakia, Egypt and Romania.

122mm M1972 (40-round) MRS — Czechoslovakia

Weight of system: 14,000kg (loaded)
Length: 8.8m (travelling)
Width: 2.5m (travelling)
Height: 2.9m (travelling)
G/clearance: 0.4m
Engine: Tatra T-930-3 V-12 air-cooled diesel developing 270hp at 2,700rpm
Speed: 75km/h
Range: 1,000km
Fuel: 500 litres
Fording: 1.4m
Gradient: 60%
Elevation of launcher: 0° to +50°
Traverse of launcher: 240° (total)
Calibre of rocket: 122mm
No of launcher tubes: 40
Length of rocket: 2.87m (long)
1.905m (short)
Weight of complete rocket: 66kg (long)
45.8kg (short)
Range: 20,380m (max, long)
11,000m (max, short)
Velocity: 690m/sec (max, long)
450m/sec (max, short)
Time to reload: 5min
Crew: 6

The 122mm M1972(40-round) multiple rocket system was developed in the late 1960s and consists of a Czechoslovak Tatra 813(8×8) truck chassis fitted with a fully armoured cab and a 122mm (40-round) rocket launcher at the very rear. The latter is identical to that used on the Soviet BM-21(40-round) multiple rocket launcher based on the Ural-375D (6×6) chassis.

Between the launcher and the cab is an additional pack of 40 rockets which can be quickly loaded into the launcher. The Tatra 813 is fitted with a central tyre pressure regulation system and some M1972s have a BZT dozer blade mounted at the front of the vehicle.

In addition to the short rocket (with a range of 11,000m) and long rocket (with a range of 20,380m), the short rocket can be fitted with an additional rocket motor which gives the rocket a range of some 17,000m.

122mm M1972 (40-round) MRS which is based on the chassis of a Tatra 813 (8×8) truck chassis.

Employment
Czechoslovakia and Germany (GDR).

147mm Rafale MRS

France

The Rafale 147mm multiple rocket system has been developed by the Société Européenne Propulsion, but has not so far been placed in production. The system is mounted on the rear of a cross country vehicle such as a Berliet GBD (6×6), with the rockets being launched from within the cab. The original launcher has a total of 18 launcher tubes (in three layers of six) but more recently the company have proposed a 30-round launcher (in three layers of ten). In the case of the 18-round launcher, all of the rockets could be fired within nine seconds with reloading being accomplished in 15min.

The rocket has a solid propellant motor and four wrap-round fins at the rear which open when the rocket leaves the launcher, air-brakes enable the

18-round Rafale 147mm MRS mounted on the rear of a Berliet GBD (6×6) truck.

range of the rocket to be varied. Two types of warhead can be fitted, anti-personnel or anti-tank. The anti-personnel warhead has 35 grenades while the anti-tank warhead has 63 shaped charged bomblets or five anti-tank mines. Rocket data:
Calibre: 147mm
Length: 3.2m
Weight: 78kg
20kg (warhead)
Range: 30,000m (max)
Velocity: 1,100m/sec (max)

Employment
Trials. Not yet in production or service.

140mm RAP-14 MRS

France

Weight of system: 4,800kg (loaded)
Length: 5.05m (travelling)
Width: 2.5m (travelling)
Height: 2.57m (travelling)
Elevation of launcher: 0° to +52°
Traverse of launcher: 360° (total)
Calibre of rocket: 140mm
Length of rocket: 2m
Span of rocket: 360mm
Weight of rocket: 54kg (complete)
19kg (HE warhead)
Range: 16,000m (max, standard rocket)
20,000m (RAP-14S) rocket

The RAP-14 MRS has been developed as a private venture by Constructions Navales et Industrielles de la Méditeranée but has not so far been placed in production although development is now complete. The launcher is mounted on a two-wheeled trailer that can be towed behind a 4,000kg truck or similar vehicle and can be prepared for firing by one man in one minute. When in the firing position the two road wheels are raised off the ground and the system is supported on three hydraulically operated jacks that are also used to level the system prior to the rockets being launched.

The launcher has a total of 22 launch rails, eg four layers of six, five, six and five rails. The rockets can be fired singly or in ripple firing, in the latter mode all 22 rockets can be fired in 10sec.

RAP-14 system.

Employment
Development complete. Ready for production.

110mm LARS

Germany, Federal Republic of

Weight of system: 15,000kg (loaded)
Length: 7.85m (travelling)
Width: 2.5m (travelling)
Height: 2.9m (travelling)
G/clearance: 0.315m
Engine: Deutz F8L 714a 8-cylinder multi-fuel developing 178hp at 2,300rpm
Speed: 73.6km/h
Range: 500km
Fuel: 150 litres
Fording: 0.86m
Gradient: 60%
Elevation of launcher: 0° to +55°
Traverse of launcher: 210°(total)
Calibre of rocket: 110mm
No of launcher tubes: 36
Length of rocket: 2.263m
Weight of rocket: 35kg (complete)
Range: 14,000m (max)
Crew: 3

110mm light artillery rocket launcher on Magirus-Deutz Jupiter (6×6) chassis.

The 110mm Light Artillery Rocket System (LARS) was developed for the German Army in the 1960s and entered service in 1970. A total of 209 systems were built and it is issued on the scale of eight launchers per division.

The system consists of a Magirus-Deutz Jupiter (6×6) truck chassis fitted with an armoured cab and a rocket launcher mounted at the rear. The rockets are launched from the cab which is provided with a roof mounted 7.62mm MG3 machine gun for anti-aircraft defence.

The launcher has a total of 36 launcher tubes in two banks of 18 with the Type 59 panoramic sight being positioned between the two banks of launchers. Before the rockets are launched, two stabilisers are lowered at the rear to provide a more stable firing platform.

The rockets are fin-stabilised and can be fired singly, rippled or one complete ripple, in the latter case all 36 rockets can be fired within 18sec. The rocket can be fitted with various types of warhead including anti-personnel, anti-tank (containing eight anti-vehicle mines), anti-tank (containing five anti-tank mines) and smoke. In 1980 the improved LARS II rocket was introduced which has a range of 20,000m compared to the original 14,000m range. It is expected that all launchers will be removed from their current Magirus-Deutz Jupiter chassis and mounted on MAN (6×6) 7,000kg truck chassis.

Employment

Germany (FRG).

Israeli MRSs

Israel has captured large quantities of multiple rocket launchers from both Egypt and Syria during both the 1967 and 1973 Middle East conflicts. Of these the Soviet 240mm(12-round) BM-24 has been reconditioned and taken into service with the Israeli Army. Rockets for the BM-24 system are now manufactured in Israel by Israel Military Industries. Each rocket has a calibre of 240mm, weighs 110.5kg at launch and 85.8kg at burnout, launch velocity is 40m/sec and velocity at impact is 270m/sec (at its max range of 10,700m). The 12-round launcher is mounted on the rear of a ZIL-157 (6×6) truck chassis which has a central tyre pressure regulation system.

Israel is also reported to use a short range rocket called Wolf. Two models have been mentioned, one with a range of 1,000m and fitted with a warhead weighing 170kg and the second with a range of 4,500m and fitted with a warhead weighing 70kg.

SAI-Ambrosini Rocket Systems — Italy

The SAI-Ambrosini Company has now taken over the responsibility for producing the hypergolic self pressurised liquid propellant rockets originally developed by the Centro Studi Transporti Missilistici. As far as it is known, none of these rockets have yet entered production. All rockets have a HE fragmentation semi-AP warhead which is detonated by an impact fuze.

Employment

Not known to be in military service.

Rocket:	*Atilla II*	*Bora*	*Mira*
Length: (travelling)	3.2m	5.3m	4m
Width: (travelling)	1.84m	1.84m	1.84m
Height: (travelling)	1.52m	1.60m	1.65m
Calibre of rocket:	82.5mm	194mm	108mm
No of launcher rails/tubes:	40	3	12
Length of rocket:	1.4m	4.7m	2.96m
Weight of rocket: (complete)	14kg	140kg	30kg
(warhead)	3kg	20kg	30kg
Range: (max)	4,500m	25,000m	15,000m
Velocity: (max at launch)	312m/sec	384m/sec	338m/sec

SNIA-Viscosa MRSs — Italy

The Defence and Aerospace Divison of SNIA-Viscosa is at present developing two multiple rocket systems as a prviate venture, these are called the FIROS 6 and the FIROS 25.

FIROS 6

This has now reached the prototype stage and consists of a 48 round launcher mounted on the rear of a FIAT 1107 AD (4×4) light vehicle. The solid propellant rockets used in the system are based on the SNIA 2inch (51mm) air-to-ground rocket. The launcher can fire the following types of rocket to a maximum range of 6,500m:

API with rocket weighting 3.80kg
HEI with the rocket weighing 3.57kg
Practice with the rocket weighing 3.57kg
Pre-fragmented with the rocket weighing 4.600kg

FIROS 25

This is currently at the prototype stage and consists of two banks of 15 122.5mm rocket launcher tubes mounted on the rear of a FIAT 6605 (6×6) truck, although the system can also be mounted on a full tracked chassis or a trailer. The weight of the rocket depends on the type of warhead fitted, the latter weigh between 17 and 39kg and max range with a 17kg warhead is 27,000m. Types of warhead available include pre-formed fragmentation, pre-formed fragmentation sub-munition.

Employment

Development. Not yet in service.

130mm Type 75 (30-round) MRL — Japan

Weight of system: 16,500kg (loaded)
Length: 5.78m (travelling)
Width: 2.8m (travelling)
Height: 2.67m (travelling)
G/clearance: 0.4m
Engine: Mitsubishi 4ZF V4 air-cooled diesel developing 300hp at 2,200rpm
Speed: 53km/h
Range: 300km
Elevation of launcher: 0° to +50°
Traverse of launcher: 100° (complete)
Calibre of rocket: 130mm
No of launcher frames: 30
Length of rocket: 1.856m
Weight of rocket: 40kg (complete)
15kg (warhead)
Range of rocket: 15,000m (max)
Crew: 3

The 130mm (30-round) Type 75 MRL was developed to meet the requirements of the Japanese Ground Self-Defence Force with Komatsu being responsible for the full tracked chassis and the Aeronautical and Space Division of the Nissan Motor Company being responsible for the rocket. The rockets, which are spin stabilised and have a solid propellant motor, can be fired singly or ripple fired. The rocket launcher is normally accompanied by a Type 75 self-propelled ground-wind measuring unit which is a derivative of the Type 73 APC.

Employment

Japan.

130mm (30-round) Type 75 multiple rocket launcher.

Type 67 Model 30 Rocket Launcher — Japan

The Type 67 Model 30 Rocket Launcher was developed by the Aeronautical and Space Division of the Nissan Motor Company Limited in the 1960s and is used only by the Japanese Ground Self-Defence Force. The complete system consists of a modified Hino (6×6) 4,000kg truck chassis with two launcher rails for the Type 67 Model 30 rockets at the rear. Before the rocket is launched three stabilisers are lowered to the ground, one either side of the vehicle and one at the rear. The rocket itself is 4.5m long, weighs 573kg at launch, has a HE warhead and a max range of between 25,000 and 30,000m. The launcher is supported by a second Hino (6×6) truck which carries six rockets and is fitted with a hydraulic crane for reloading purposes.

Type 67 Model 30 rocket launcher in travelling order.

Type 67 Model 30 rocket launcher from the rear with rocket resupply vehicle to its right.

South Korean MRSs

South Korea has developed a 28-round multiple rocket launcher which is mounted on the rear of American supplied M809 series (6×6) 5,000kg truck. The launcher has four layers of eight launcher tubes and the rockets are thought to have a calibre of between 120mm and 150mm. The South Korean Army has modified the American supplied Nike-Hercules SAM for use in the surface-to-surface role.

Employment

South Korea.

140mm WP-8 (8-round) MRS — Poland

Weight of system: 687.6kg (loaded)
370kg (unloaded)
Length: 3.294m (travelling)
Width: 1.634m (travelling)
Height: 1.2m (travelling)
G/Clearance: 0.254m
Track: 1.4m
Elevation of launcher: 0° to +45°
Traverse of launcher: 30° (total)
Calibre of rocket: 140.4mm

140mm (8-round) WP-8 MRS being towed by GAZ-69 (4×4) vehicle.

No of launcher tubes: 8
Length of rocket: 1.085m
Weight of rocket: 39.7kg (complete)
Range: 9,810m (max)
Velocity: 400m/sec (max)
Time to reload: 2min
Crew: 5

The 140mm (8-round) WP-8 multiple rocket system was developed for use by Polish airborne units and entered service in the 1960s. The launcher has eight tubes in two layers of four and is mounted on a split trail carriage. The launcher fires the same spin-stabilised rockets as the Soviet 140mm (16-round) RPU-14 (towed), 140mm (16-round) BM-14-16 (on ZIL-151 or ZIL-131 chassis) and the 140mm (17-round) BM-14-17 (on GAZ-63A) multiple rocket launchers.

Employment

Poland.

Spanish MRSs

At least four types of multiple rocket system have been developed in Spain and brief details of these are given below.

D-10 System This has a 10-round launcher (in two layers of five) mounted on the rear of a Barreiros Panter (6×6) truck chassis. The system fires 300mm D3 rockets which weigh 247.5kg to a max range of 17,000m.
E-20 System This trailer mounted system has four rows of five rockets and fires an 108mm E2B rocket weighing 16.4kg to a max range of 7,500m.
E-32 System This has a 32-round launcher (in four layers of eight) mounted on the rear of a 4×4 truck chassis and fires and 108mm R6B2 rocket weighing 19.4kg to a max range of 10,000m.
E-21 System This has a 21-round launcher (in three layers of seven) mounted on the rear of a Barreiros Panter (6×6) truck chassis and fires a 216mm E3 rocket weighing 10kg to a max range of 14,500m.

Employment

Spain.

300mm D-10 MRSs mounted on rear of Barreiros Panter (6×6) truck chassis

E-20 trailer mounted 108mm MRL.

81mm Oerlikon RWK-007 MRS — Switzerland

The 81mm RWK-007 MRS has been developed as a private venture by Machine Tool Works Oerlikon-Bührle Limited of Zurich for the export market. The launcher has two parallel banks of 15 tubes each with the complete launcher having an elevation of +50°, a depression of −10° and a total traverse of 360°, both elevation and traverse are manual. The rockets are aimed using an optical sight with a magnification of ×7 and a 9° field of view. The launcher has loaded weight of about 1500kg and can be mounted on the roof of AFVs such as the American M113, MOWAG Tornado MICV and members of the MOWAG Piranha range of wheeled AFVs.

The launcher can fire both DIRA and SNORA 81mm folding fin rockets, the former have a maximum range of 8,500m. The DORA rockets can be fitted with the following types of warhead: fragmentation explosive, hollow charge, illuminating, marker and practice, while the SNORA rockets can be fitted with either fragmentation explosive (three different weights), hollow charge or practice.

Employment

Oerlikon have released no details of the employment of the RWK-007 multiple rocket system.

81mm Oerlikon RWK-007 MRS mounted on MOWAG Piranha 8×8 amphibious vehicle.

250mm BM-25 (6-round) MRS — USSR

Weight of system: 18,145kg (loaded)
Length: 9.815m (travelling)
Width: 2.7m (travelling)
Height: 3.5m (travelling)
G/clearance: 0.38m

Engine: YaMZ-206B 6-cylinder water-cooled diesel developing 205hp at 2,000rpm
Speed: 55km/h
Range: 530km
Fuel capacity: 450 litres

Fording: 1m
Gradient: 60%
Elevation of launcher: 0° to +55°
Traverse of launcher: 6° (total)
Calibre of rocket: 250mm
No of launcher frames: 6
Length of rocket: 5.822m
Weight of rocket: 455kg
Range: 30,000m (max)
Time to reload: 10-20min
Crew: 8 to 12

The 250mm (6-round) BM-25 is the largest MRL to have entered service with the Soviet Army in the postwar period. It is no longer in front line service but some are probably held in reserve. The six-round launcher is mounted on the rear of a KrAZ-214 (6×6) cross-country truck which has a central tyre pressure regulation system.

Employment

Soviet Union (reserve).

250mm (6-round) BM-25 MRSs on parade in Moscow.

240mm BM-24 (12-round) MRS USSR

Weight of system: 8,680kg (loaded)
Length: 6.705m (travelling)
Width: 2.315m (travelling)
Height: 2.910m (travelling)
G/clearance: 0.31m
Engine: ZIL-157 6-cylinder water-cooled petrol developing 109hp at 2,800rpm
Speed: 65km/h
Range: 430km
Fuel capacity: 215 litres
Fording: 0.85m
Gradient: 55%
Elevation of launcher: 0° to +65°
Traverse of launcher: 140° (total)
Calibre of rocket: 240mm
No of launcher frames: 12
Length of rocket: 1.29m (long)
1.225m (short)
Weight of rocket: 109kg (long)
112kg (short)
Range: 10,200m (max, long)
6,575m (max, short)
Time to reload: 3-4min
Crew: 6

240mm (12-round) BM-24 MRSs of the Israeli Army.

The 240mm (12-round) BM-24 MRS entered service with the Soviet Army in the 1950s and fires the same rockets as the 240mm (12-round) BM-24T system which is based on the AT-S medium tracked artillery tractor.

When originally introduced the system was mounted on the rear of a ZIL-151 (6×6) truck chassis, this was later replaced by the improved ZIL-157 (6×6) truck chassis which has single rear wheels and a central tyre pressure regulation system. The BM-24 has been replaced in many Soviet front line units by the 122mm (40-round) BM-21 multiple rocket system.

The BM-24 and BM-24T can fire two types of spin-stabilised rocket, one with a range of 10,200m and a light warhead and the second with a range of 6,575m and a heavier warhead.

Employment

Algeria, Egypt, Germany (GDR), Israel (rockets manufactured by Israel Military Industries), North Korea (unconfirmed), Poland, Syria and the USSR.

240mm BM-24T (12-round) MRS — USSR

Weight of system: 15,240kg (loaded)
Length: 5.87m (travelling)
Width: 2.57m (travelling)
Height: 3.1m (travelling)
G/clearance: 0.4m
Engine: V-54-T V-12 water-cooled diesel developing 250hp at 1,500rpm
Speed: 35km/h
Range: 380km
Fuel capacity: 420 litres
Fording: 1m
Gradient: 50%
Elevation of launcher: 0° to +45°
Traverse of launcher: 210° (total)
Calibre of rocket: 240mm
No of launcher tubes: 12
Length of rocket: 1.29m (long)
1.225m (short)
Weight of rocket: 109kg (long)
112kg (short)
Range of rocket: 10,200m (max, long)
6,575m (max, short)
Time to reload: 3-4 minutes
Crew: 6

The 240mm (12-round) BM-24T multiple rocket launcher fires the same rockets as the 240mm (12-round) BM-24 system which is based on a ZIL-157 (6×6) truck chassis. The BM-24T is mounted on the rear of an AT-S Medium Tracked Artillery Tractor and has a tube type launcher rather than the frame type launcher as used with the BM-24 system. The BM-24T is no longer in front line service with the USSR but they are probably held in reserve, its replacement was the 122mm (40-round) BM-21 multiple rocket system.

Employment

USSR (reserve).

200mm BMD-20 (4-round) MRS — USSR

Data: System based on ZIL-151 truck chassis
Weight of system: 8,700kg (loaded)
Length: 7.2m (travelling)
Width: 2.3m (travelling)
Height: 2.85m
G/clearance: 0.265m
Engine: ZIL-121 6-cylinder water-cooled petrol developing 92hp
Speed: 60km/h
Range: 600km
Fuel capacity: 300 litres
Fording: 0.8m
Gradient: 50%
Elevation of launcher: +9° to +60°
Traverse of launcher: 20° (total)

200mm (4-round) BMD-20 MRSs parade through Red Square, Moscow in 1949.

Calibre of rocket: 200mm
No of launcher frames: 4
Length of rocket: 3.11m
Weight of rocket: 91.4kg
Range: 20,000m (max)
Time to reload: 6-10min
Crew: 6

The 200mm (4-round) BMD-20 MRS was first deployed with the Soviet Army in the 1950s and mounted on a ZIL-151 (6×6) truck chassis, this was later replaced by the improved ZIL-157 (6×6) truck chassis which has single rather than dual rear wheels and a central tyre pressure regulation system. The BMD-20 is longer in front line service with the USSR but they are probably held in reserve.

Employment

USSR reserve and Korea (North).

140mm BM-14-17 (17-round) MRS — USSR

Weight of system: 5,323kg (loaded)
Length: 5.41m (travelling)
Width: 1.985m (travelling)
Height: 2.245m (travelling
G/clearance: 0.27m
Engine: GAZ-51A 6-cylinder water-cooled petrol developing 70hp at 2,800rpm
Speed: 65km/h
Range: 650km
Fuel capacity: 195 litres
Fording: 0.8m
Gradient: 50%
Elevation of launcher: 0° to +50°
Traverse of launcher: 180° (total)
Calibre of rocket: 140.4mm
No of launcher tubes: 17
Length of rocket: 1.085m
Weight of rocket: 39.7kg (complete) 18.4kg (warhead)
Range: 9,810m (max)
Velocity: 400m/sec (max)
Time to reload: 2min
Crew: 6

The 140mm (17-round) BM-14-17 MRS entered service with the Soviet Army in the 1950s and fires the same rockets as the Polish 140mm (8-round) WP-8 (towed), Soviet 140mm (16-round) RPU-14 (towed) and 140mm (16-round) BM-14-16 (on ZIL-151 or ZIL-131 chassis) multiple rocket systems. In the case of the BM-14-17 the system is mounted on the rear of a modified GAZ-63A (4×4) truck chassis. The system is only in limited use today.

Employment

Poland and the USSR.

140mm (17-round) BM-14-17 MRSs on GAZ-63A (4×4) truck chassis.

140mm BM-14-16 (16-round) MRS USSR

Weight of system: 8,200kg (loaded)
Length: 6.92m (travelling)
Width: 2.3m (travelling)
Height: 2.265m (travelling)
G/clearance: 0.265m
Engine: ZIL-121 6-cylinder water-cooled petrol developing 92hp
Speed: 60km/h
Range: 600km
Fuel capacity: 300 litres
Fording: 0.8m
Gradient: 50%
Elevation of launcher: 0° to +50°
Traverse of launcher: 140° (total)
Calibre of rocket: 140.4mm
No of launcher tubes: 16
Length of rocket: 1.085m
Weight of rocket: 39.7kg
18.4kg (warhead)
Range: 9,810m (max)
Velocity: 400m/sec (max)
Time to reload: 3min
Crew: 7

The 140mm (16-round) BM-14-16 MRS was introduced into the Soviet Army in the early 1950s and fires the same spin-stabilised rockets as the Polish 140mm (8-round) WP-8 (towed), Soviet 140mm (16-round) RPU-14 (towed) and 140mm (17-round) BM-14-17 (on GAZ-63A 4×4 chassis) multiple rocket systems. When originally introduced the system was mounted on the rear of a ZIL-151 (6×6) truck chassis but some have recently been mounted on the rear of the ZIL-131 (6×6) truck chassis which has single rear wheels and a central tyre pressure regulation system. The system has been replaced in many Soviet units by the 122mm BM-21 (40 round) and 122mm M1972 (40-round) MRSs.

Employment

Algeria, China, Egypt, Korea (North), Poland, Syria, USSR and Vietnam.

140mm (16-round) BM-14-16 MRS on ZIL-151 (6×6) truck chassis.

140mm RPU-14 (16-round) MRS USSR

Weight of system: 1,200kg (loaded)
Length: 4.036m (travelling)
Width: 1.80m (travelling)
Height: 1.60m (travelling)
G/clearance: 0.315m
Track: 1.45m
Elevation of launcher: 0° to +45°
Traverse of launcher: 30° (total)
Calibre of rocket: 140.4mm
No of launcher tubes: 16
Length of rocket: 1.085m
Weight of rocket: 39.7kg
18.4kg (warhead)
Range: 9,810m (max)
Velocity: 400m/sec (max)
Time to reload: 4min
Crew: 5
Towing vehicle: GAZ-66 (4×4) truck

The 140mm (16-round) RPU-14 MRS has been designed for use with Soviet Airborne Rifle Divisions and is issued on the scale of 18 per division. The system fires the same spin stabilised rockets as the Polish 140mm (8-round) WP-8 (towed), Soviet 140mm (16-round) BM-14-16 (on ZIL-151 or ZIL-131 6×6 truck chassis) and the 140mm (17-round) BM-14-17 (on GAZ-63A 4×4 chassis) multiple rocket system.

Employment
USSR.

140mm (16-round) RPU-14 MRSs being decontaminated by their crews.

132mm BM-13-16 (16-round) MRS — USSR

Weight of system: 6,432kg (loaded)
Length: 7.5m (travelling)
Width: 2.3m (travelling)
Height: 3.19m (travelling)
G/clearance: 0.265m
Engine: ZIL-121 6-cylinder water-cooled petrol developing 92hp
Speed: 60km/h
Range: 600km
Fuel capacity: 300 litres
Fording: 0.8m
Gradient: 50%
Elevation of launcher: +15° to +45°
Traverse of launcher: 20° (total)
Calibre of rocket: 132mm
No of launcher rails: 8(each with two rockets)
Length of rocket: 1.473m
Weight of rocket: 42.5kg
Range: 9,000m (max)
Velocity: 350m/sec (max)
Time to reload: 5-10min
Crew: 6

The 132mm BM-13-16 (16-round) MRS was introduced into the Soviet Army during World War 2 and was originally mounted on the rear of a Soviet or American 6×6 truck chassis. In the postwar period they were removed from their original chassis and

132mm BM-13-16 (16-round) MRS on rear of ZIL-151 (6×6) truck chassis.

fitted on to ZIL-151 (6×6) truck chassis. An unusual feature is that each of the eight I-shaped launcher rails have two rockets, one on the top and another on the bottom. The system is no longer in front line service with any members of the Warsaw Pact, although some countries may use it for training purposes.

Employment

Afghanistan and China.

122mm BM-21 (16-round) MRS — USSR

Weight of system: 11,500kg (loaded)
Length: 7.35m (travelling)
Width: 2.69m (travelling)
Height: 2.85m (travelling)
G/clearance: 0.41m
Engine: ZIL-375 V-8 water-cooled petrol developing 180hp at 3,200rpm
Speed: 75km/h
Range: 405km
Fuel capacity: 360 litres
Fording: 1m
Gradient: 60%
Elevation of launcher: 0° to +50°
Traverse of launcher: 240° (total)
Calibre of rocket: 122mm
No of launcher tubes: 40
Length of rocket: 2.87m (long)
1.905m (short)
Weight of rocket: 66kg (complete, long)
45.8kg (complete, short)
Range: 20,380m (max, long)
11,000m (max, short)
Velocity: 690m/sec (max, long)
450m/sec (max, short)
Time to reload: 10min
Crew: 6

The 122mm BM-21 (40-round) MRS was developed in the 1960s and has now become the standard system of the Warsaw Pact, it has also been exported in large numbers and used operationally in both Africa and the Middle East. It has now been supplemented in some Warsaw Pact countries by the 122mm M1972 (40-round) system which consists of a Tatra 813 (8×8) truck chassis with an armoured cab and the same rocket launcher as that fitted to the BM-21 mounted at the very rear. To the rear of the cab is an additional pack of 40-rockets which can be quickly loaded into the launcher. Details of the M1972 are given in this section under Czechoslovakia.

The BM-21 is based on the chassis of the Ural-375D (6×6) truck which has a central tyre pressure regulation system and is issued on the scale of 18 per division.

In addition to the short rocket (with a range of 11,000m) and the long rocket (with a range of 20,380m), the short rocket can be fitted with an additional rocket motor which gives the rocket a range of some 17,000m.

Employment

Angola, Bulgaria, Egypt, Ethiopia, Germany (GDR), Hungary, Iran, Iraq, Israel, North Korea, Mozambique, Poland, Syria, USSR, Vietnam and Yemen (South).

122mm BM-21 (40-round) MRSs on parade in Red Square, Moscow.

Multiple Launch Rocket System

USA

Weight of system: 22,680kg (loaded)
Length: 6.97m (travelling)
Width: 2.97m (travelling)
Height: 2.59m (travelling)
G/clearance: 0.43m
Engine: Cummins VTA-903 turbocharged diesel developing 500hp at 2,400rpm
Speed: 64km/h
Range: 483km
Fuel capacity: 617 litres
Fording: 1.02m
Gradient: 60%
Calibre of rocket: 227mm
No of launcher tubes: 12
Length of rocket: 4m approx
Range: at least 30,000m
Crew: 3

In 1976 the Army Missile Research and Development Command studied the possibility of developing a long range MRS to help offset the growing number of Warsaw Pact artillery weapons. In March 1976 five companies were awarded contracts for the concept definition stage of the system which was then given the name of the General Support Rocket System.

After studying the five proposals submitted, Boeing Aerospace and the Vought Corporation were each awarded a contract for the validation phase of this competition. Under the terms of this contract, each company built three launch systems and some 150 rockets. In April 1980 Vought were declared the winner of the MLRS competition and awarded a contract for final development. First production systems will be completed in 1982.

In 1979, France, Germany (FGR) and the UK signed a Memorandum of Understanding with the USA for the joint production of the system which then became known as the Multiple Launch Rocket System. The final details have yet to be decided but to production lines will probably be established, one in the USA and one in Europe.

The rockets are transported and launched from a six-round pod, with each system having two such pods. The rockets can be launched from within the cab singly or ripple fired, and once the two pods have been used two new pods can be quickly loaded. The rockets can be fitted with a number of different warheads including dual purpose anti-material/anti-personnel, scatterable anti-tank mine and a terminal homing anti-tank munition.

The MLRS is mounted on the rear of a full tracked carrier that is based on the chassis of the FMC developed XM2 Infantry Fighting Vehicle and the XM3 Cavalry Fight Vehicle.

Employment

To enter service in 1983.

Vought Multiple Launch Rocket System launching a rocket from its left pod.

115mm M91 MRL

USA

The M91 (development designation T145) was designed to fire the M55 chemical rocket with either a VX or GB warhead. The GB rocket is 1.981m long and weighs 24.9kg while the VX rocket is 1.981m long and weighs 25.40kg. For training purposes the M60 dummy and M61 training rockets are provided. It has a total of 45 tubes arranged in five rows of nine each, the 45 rockets could be ripple fired in some 15 seconds. Emplacement time, including loading the 45 rockets was 30min.

The M91 is transported (ie carried in) a 6×6 M35 truck, the small wheels of the M91 being designed for moving the launcher only short distances. The mounting is provided with jacks for levelling purposes. The M91 can also be fired whilst still in the truck and is air-transportable. Brief data is:

Weight of launcher: 544kg (w/o rockets)
1,912kg (with rockets)
Length: 3.86m
Width: 2.971m
Height: 1.701m
Traverse: 360° (on wheels)
10° (left or right)
Elevation: +1° to +60°
Range: 10,972m

Employment

In service with the US Army only.

The M91 115mm MRL firing.

128mm YMRL32 MRS

Yugoslavia

Data: Provisional and relates to system based on FAP 2220BDS 6×6 truck chassis
Weight of system: 13,000kg (loaded)
Length: 7.75m (travelling)
Width: 2.46m (travelling)
Height: 2.97m (travelling)
Speed: 60km/h
Range: 700km
Fuel capacity: model 2F/002A 6-cylinder water-cooled diesel developing 200hp
Calibre of rocket: 128mm
No of launcher tubes: 32
Range: 18,000m (max new rocket)
9,600m (max old rocket)
Time to reload: 2min
Crew: 3

This system was developed in Yugoslavia in the early 1970s and has been given the designation YMRL 32 128mm in the West, with YMRL standing for Yugoslav Multiple Rocket Launcher.

In concept it is similar to the Czechoslovak 122mm (40-round) M1972 system but the Yugolsav model is unarmoured. When originally introduced the system was mounted on the FAP 2220BDS (6×6) truck chassis, recently the system has been seen mounted on the chassis of the FAP 2020BS (6×6) truck. This has larger tyres with a central tyre pressure regulation system and therefore has improved cross-country performance over the earlier vehicle.

The 32-round rocket launcher is mounted at the very rear of the chassis with a pack of 32 rounds being mounted to the immediate rear of the cab ready for rapid reloading. Some vehicles have a 7.92mm AA MG mounted on the roof of the cab. The system fires a new long range (18,000m) rocket in addition to the older rocket (range 9,600m) which is used by the towed 128mm (32-round) M-63 MRS.

Employment

Yugoslavia.

128mm M-63 (32-round) MRS — Yugoslavia

Weight of system: 2,134kg (loaded)
Length: 3.595m (travelling)
Width: 1.634m (travelling)
Height: 1.278m (travelling)
G/clearance: 0.327m
Elevation of launcher: 0° to +48°
Traverse of launcher: 30° (total)
Calibre of rocket: 128mm
No of launcher tubes: 32
Length of rocket: 0.8m
Weight of rocket: 23kg
Range: 9,600m (max)
Velocity: 420m/sec (max)
Time to reload: 3min
Crew: 5
Towing vehicle: TAM 1500 (4×4) truck

The 128mm (32-round) M-63 MRL entered service with the Yugoslav Army in the 1960s and has four layers of eight launcher tubes, with the carriage being of the split trail type, The rockets used in the system can also be fired by the YMRL 32 128mm truck-mounted MRS, this fires a new rocket to a max range of 18,000m compared to the 9,600m of the M-63 towed system.

Employment

Yugoslavia.

128mm (32-round) M-63 MRS.

Part Three: Artillery Fire Control Systems and Other Aids

MILIPAC Artillery Computer — Canada

The MILIPAC artillery computer has been developed from 1976 by the Computing Devices Company of Canada under contract to the United States Army. At a later date the Canadian Government ordered three prototypes for evaluation and these were delivered in 1979, a fourth prototype has been delivered to Spain.

The artillery computer has been designed to calculate firing data for a battery of six 105mm or 155mm guns as weli as being used in the field and survey roles. It is man portable but is normally mounted in the tracked or wheeled vehicle and deployed at battery level.

Data is entered manually using a keyboard, computer used in the MILIPAC is a Control Data Model 469C. The liquid crystal display has a capacity of 10 × 48 character lines and provides both input and output data as well as cue messages to assist the operator.

Employment

Trials with Canadian Armed Forces and Spain.

DR513 Muzzle Velocity Measurement Equipment — Denmark

The DR513 was built by Radartronic A/S and is based on the doppler principle. It can measure muzzle velocities ranging from 50 to 2,000m/sec, accurate to 0.1%.

The basic equipment for measuring the muzzle velocity consists of: the antenna unit; the data unit; the test unit. The doppler signal from the antenna unit is directly proportional to the velocity and is fed to an advanced electronic data unit where the results can be read out on four display tubes. The figure on the display is the time the projectile has spent travelling a distance of 2m measured 30 and 40m in front of the muzzle. The corresponding velocity can easily be found in tables. The correction factor for the geometrical error introduced by offsetting the antenna from the line of fire can also be found in the tables. The test unit is a simulator providing a fast control on function and transmitter frequency.

The antenna unit is designed to withstand a very high degree of shock and vibration and can be placed next to or even on the gun, to provide a minimum geometrical error. The use of the DR513 makes it possible to measure the velocity of all sizes of projectiles. The measuring range is about 5,000 times the calibre.

The DR513 basic equipment can be extended with various units for analog and digital measurements throughout the trajectory. The analog data unit together with an ultra violet paper recorder give a curve showing the velocity as a function of time. The ballistic trajectory analyser (BATRAN) system together with a tape puncher give results on a punched tape ready for data handling in a computer. The results can also be printed directly on a digital printer.

The DR513 has been succeeded in production by the DR810. This was developed by B&W Elektronik (previously Radartronic and then Dannebrog Elektronic) but this has been licensed to Lear Siegler Incorporated of Santa Monica, California, and is now marketed through their division. Full details of the DR810 are therefore given under the USA.

Employment

Denmark, Finland, Germany (FRG), Holland, Italy, Malaysia, Mexico, Norway, Pakistan, Singapore, Sweden, Switzerland, UK and Yugoslavia.

155mm gun at point of firing. The antenna unit is to the left of the wheels.

TM12 Artillery Rangefinder/Sight

France

The TM12 Artillery Rangefinder/Sight has been developed by Sopelem and CILAS (Compagnie Industrielle des Lasers), under a contract from DTAT/SEFT and is now in production for the French Army. It can be used for both measuring elevation and azimuth angles by optical aiming and distance by laser rangefinding.
The TM12 consists of the following sub-assemblies:
Optical Housing — optics and elevation control
Goniometer Plate — goniometer, azimuth control, bubble levels and tripod.
Rangefinder Casing — laser emission, laser reception, electronic clock, readout and controls, power supplies.
The laser rangefinder is accurate to ±5m, minimum range is 150m and maximum range is 20,000m. The min ranging gate is adjustable between 150 and 5,000m. Electrical power is supplied by a NiCd battery.

The observation and aiming sight has a magnification of ×8 and can be traversed through 360°, elevation range is ⁻20° to +40°. The complete sight weighs 12kg, the battery 1kg and the carrying case 2kg.

Employment

In service with the French Army.

The TM12 artillery rangefinder/sight on its tripod and complete with battery.

TPV89 Laser Rangefinder

France

The TPV89 handheld laser rangefinder has been developed by Sopelem and CILAS (Compagnie Industrielle des Lasers) to enable forward observers to obtain fast and accurate measurement of distance to the target. All the operator has to do after switching on is to centre the cross-hairs on the target and trigger the laser by means of a push-button located on the top. The distance in metres is instantantaneously displayed in the eyepiece.

The rangefinder is accurate to ±10m whatever the distance of the target within the range limits. An optical device provides absolute protection of the operator's eye during the emission of the laser. The TPV89 also has a safety device to prevent accidental triggering, a multiple echo indicator and a minimum ranging gate with a pilot lamp that lights when this is in use.

The monocular sight has a magnification of ×6 and a 7° field of view and the 12V NiCd battery has sufficient power for at least 600 shots. Min range is

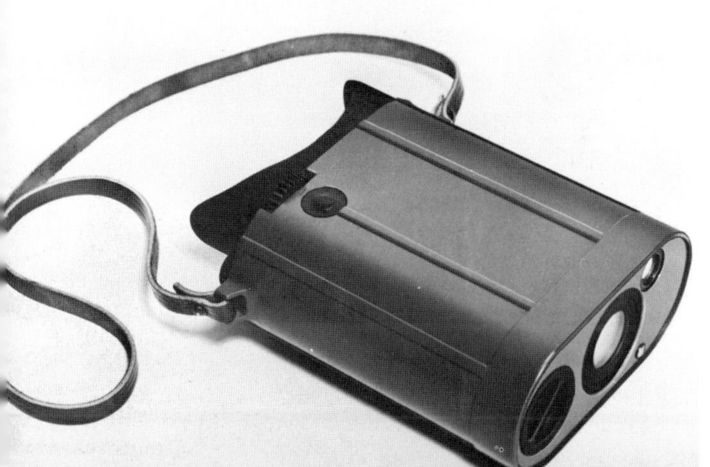

The TPV89 handheld laser rangefinder.

150m and max range is 9,990m, weight with NiCd battery is 1.9kg and weight with lithium primary battery is 1.7kg. Optional accessories include a tripod with elevation and azimuth control and a charger for the NiCd battery.

Employment

In production.

CAC 101 Field Artillery Computer — France

The Thomson-Brandt CAC 101 artillery computer has been developed as a private venture to calculate firing data for 105mm and 155mm guns and howitzers as well as heavy mortars of the type manufacturerd by the company.

The CAC 101 is 370 × 355 × 213mm and weighs 15kg, the battery supplies sufficient power for two hours of continuous operation. The computer memory can store data for six gun positions, ten target locations, ten observer positions, ten protected areas as well as meteorological data (analytical meteorological message STANAG 4082 up to line 21). Plug in modules contain the data for different types of weapon and projectiles.

The computer can compute fire data for the base weapon as well as other weapons in the battery, permanent self-checking of the system is performed by means of a built-in test programme.

Data is entered into the computer using a keyboard display according to a dialogue between the operator and the computer. This dialogue has been made possible by the alpha-numerical panel. The computer automatically gives warning if an impossible or dangerous fire mission is given to it.

In addition to calculating fire data, the computer can calculate the ballistic coordinates of a given round during fire adjustment, computation of residual correction at the end of registration, computation of correction for muzzle velocity as well as various topographic computations.

Employment

in service with undisclosed countries.

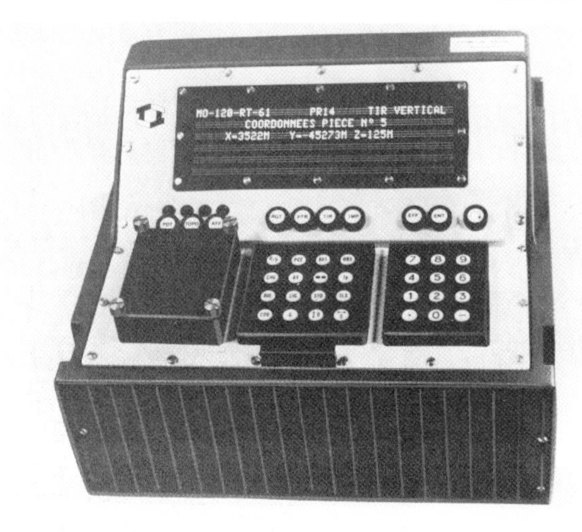

Thomson-Brandt CAC 101 field artillery computer.

EMD Sirocco Meteorological Radar System — France

The Sirocco meteorological radar system has been developed by Electronique Marcel Dassault for the French Army under the direction of the SEFT. The system has been designed to provide artillery batteries with accurate meteorological data for gun laying and consists of a two-axis monopulse radar, built in telemetry receiver and a special purpose computer.

The radar is mounted on a two wheeled trailer while the air conditioned SH17 shelter, which is mounted on the rear of a Berliet GBC 8KT (6×6)

The Sirocco radar deployed in action.

truck (or any similar vehicle), houses the operator, the radar and telemetry operating console with its scope, a radio-telegraphy system and built in stowage for the radar reflectors. In addition to carrying the radar, the trailer also carries the radar transmitter/receiver, the telemetry receiver and two generators.

The radar automatically measures atmospheric temperatures, wind speed and direction by tracking sensors carried aloft by ballons. The radar can track balloons out to ranges in excess of 130km and the computer can determine wind speed to within one knot. The station prints out in real time on a teleprinter either the standard STANAG 4082 message used by gunnery computers or the ballistic STANAG 4061 message for field artillery units.

Employment

In 1979 EMD announced that they had received orders for over 30 Sirocco systems from both the French Army and overseas customers, the latter includes Morocco.

ATILA Artillery Automation System — France

The ATILA artillery automation system has been designed by CIMSA (Compagnie d'Information Militaire Spatiale et Aeronautique) to meet the requirements of the French Army for a system capable of computing data for artillery, transmission of data from the forward observer to the command post and then on to the guns, and finally management of tactical data.

Main components of the system are: (a) A forward observer terminal for the coding and transmission of fire requests and intelligence messages, the observer has a TM12 laser rangefinder for which there is a separate entry in this section; (b) Fire command and control post located at the command post of the regimental commander who is in charge of fire request co-ordination, threat evaluation, assignment of fire units and the calculation of firing data when this is not accomplished at the batteries; (c) Fire execution centre at the batteries which calculates the firing data and also receives fire requests when the battery is decentralised; (d) Terminal for the liaison officer in charge of co-ordinating the requirements of the supported unit and the resources of the fire units; (e) Remote display terminal at each gun which receives the firing data.

The ATILA artillery automation system can also be used in conjunction with the Sirocco meteorological system (separate entry), muzzle velocity measuring equipment on each artillery weapon, RATAC battlefield radar (separate entry), vehicle navigation system and a complete command and control system.

Employment

In production for the French Army and one undisclosed overseas army.

Falke Artillery Computer System — Germany, Federal Republic of

The Falke artillery computer system has been developed by the radio and radar division of AEG-Telefunken to meet the requirements of the German Army.

The central unit of the Falke system is the freely programmable digital computer, the TR84, which has been designed specifically for military operations under a wide range of operational environments. The other two components are the operator control panel and the program-loading device. The total weight of Falke, complete with the operator panel, interconnecting panel and the program-loading device is 61.3kg. Ballistic programmes are available for various weapons ranging in calibre from 80mm to 203mm as well as missiles. A special data output device, called the Visual Indicator OA24, can be coupled directly to the computer or used away from the computer, eg by a forward observer, with the data being transmitted directly to the computer by a radio link. The computer can be run from a 24V power supply and features a word length of 18 bits and a core memory with a capacity of 16K words.

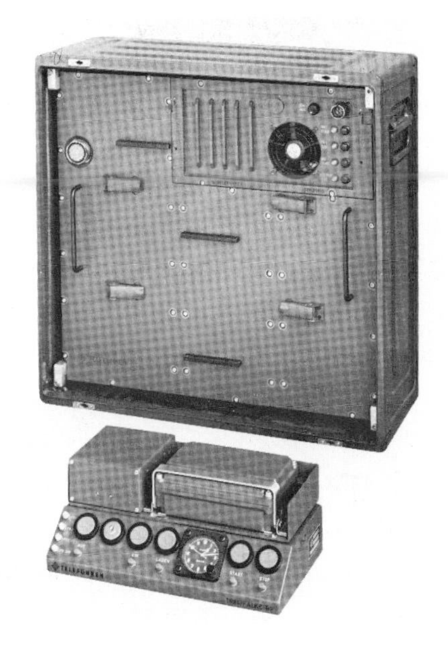

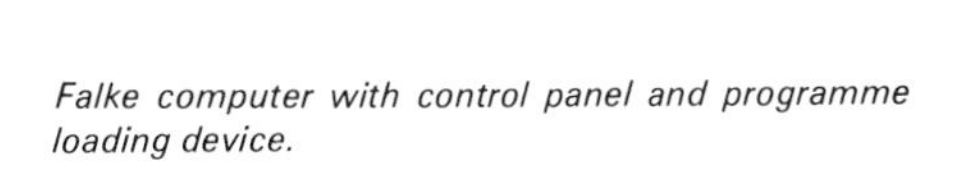

Falke computer with control panel and programme loading device.

The Falke system has the following applications: fire direction (gun and rocket), meteorological data evaluation, reconnaissance with sound ranging, reconnaissance with flash ranging, surveying procedures, and the establishment of specific meteorological messages.

Employment

In service with the German (FRG) Army.

RATAC Battlefield Radar — International

The RATAC (Radar de tir pour l'artillerie de campagne) was originally developed by the French company of Laboratorie Central de Télécommunications in the 1960s. Production was subsequently undertaken in France, West Germany (Standard Elektrik Lorenz) and the United States (Gilfillan). In the French Army is mounted on an AMX-10P or AMX VCI infantry fighting vehicle while in the German Army it is mounted on a modified M113 APC.

The radar can be used for a variety of roles including the detection, acquisition, identification, location and tracking of targets such as tanks, vehicles and even low flying helicopters. Max range against vehicle targets is reported to be 20,000m while max range against troops is about 10,000m. In addition the system can be used for other roles such as ground surveillance and artillery direction. Features of the RATAC include automatic tracking facilities, the capability to transmit data to artillery units and as an option a plotting board unit can be fitted.

RATAC battlefield radar mounted on the roof of a French AMX-10P infantry fighting vehicle.

Employment

France, Germany (FGR) and the USA (AN/TPS-58).

DAVID Field Artillery Computer — Israel

The DAVID field artillery computer has been developed by the RAFAEL Armament Development Authority of Haifa to perform all of the calculations at the battery level, replacing all the traditional tables and charts. DAVID provides accurate results in a fraction of the time required for conventional manual calculations, for example for fire data it requires 5-7sec; for chart data 2sec, for corrections 10-15sec and for self-test 3sec. It weighs only 23kg and is adaptable to any gun or mortar and can handle up to six guns and two fire missions simultaneously. Up to 28 targets and/or alternate positions can be stored for quick reference and up to nine corrections can be stored for ease of operation.

DAVID is composed of a main unit, which includes keyboard and display and a plug in memory module and is simple to maintain and operate. Its memory size is 24K, power supply 20-35V dc (nominal 24V), with a 25W power consumption in the operating mode and 8W power consumption in the stand-by mode. For short periods, DAVID can operate by means of an internal rechargeable battery, and the system is capable of operating in a temperature range of −10° to +52°C.

A companion piece of equipment is the Gun Display System (GDS) which immediately transfers the essential data between DAVID at the fire direc-

DAVID field artillery computer.

tion center (FDC) and the guns. There is also a meteorological version of the DAVID which performs all the calculations normally carried out by the meteorological crew. This replaces most of the charts, books and rules normally used and can compile the meteorological message in a very short time.

Employment

Israel.

ELTA EL/M-2106 Point Defence Alert Radar

Israel

DNS

The EL/M-2106 point defence alert radar has been designed by ELTA Electronic Industries Limited to provide light anti-aircraft guns and man-portable SAMs with early warning of the approach of enemy aircraft.

It has been designed to be disassembled into light and easily transportable units with each unit being carried by a single soldier. The radar can be deployed in less than 10 minutes and one operator is sufficient for both set up and operation.

The radar consists of three main components, transceiver (operating in the L-band and weighing 26kg), antenna (weighing 55kg including stand and pedestal) and the synthetic display with sweep memory (weight 2kg). Power is supplied by standard 24V batteries.

The radar continuously sweeps through 360° and can be operated remotely from up to 100m away. At normal attacking speeds, with a detection range of 16km, the alarm time exceeds one minute, sufficient for the anti-aircraft defences to be alerted and the gunners pointing their weapons in the direction of the threat and expect the attacking aircraft rather than being surprised by it.

The synthetic display has 36 circles of LEDs, with azimuth presentation at 10° intervals and range presentation at 1km intervals (max range 8km) or 2km intervals (max range 16km).

Employment

No details available.

LPD/20 Pulse Doppler Search Radar

Italy

Dows

The LPD/20 Search Radar has been designed and built by Contraves Italiana SpA of Rome. Two models have been developed, one of these being mounted on a two wheeled trailer and the other on a 4×4 cross-country truck. Their operational characteristics are identical.

The LPD/20 can be used in a number of ways ie integrated with the Super-Fledermaus FCS, used simply as a mobile radar station or for alerting conventional anti-aircraft guns or man portable SAM's. When integrated with the Super-Fledermaus a data transmission system is required and in this configuration, one sole PPI, that of the fire-control system, covers all functions so that the search radar can be operated unattended and a single operator is capable of controlling the entire radar system, both in respect of searching and tracking. Basic data of the trailer mounted LPD/20 is: length, 4.01m; width, 2.02m; height, 2.8m, weight, 2,200kg.

The trailer mounted LPD/20 consists of the following main components: (a) A two wheeled trailer; (b) Radar scanner consisting of the antenna transceiver unit, modulator and signal processor. The antenna rotates at 30 or 60rpm; (c) The cabin which mounts the driving mechanism for the antenna, the high power supply unit, and on the rear of the trailer, the power supply panel and the operational panel (d) The 50/400Hz converter is located between the cabin and the towbar and this supplies 400Hz power to all circuits (e) PPI display and console. This is connected by cables to the radar and enables the operator to control it remotely.

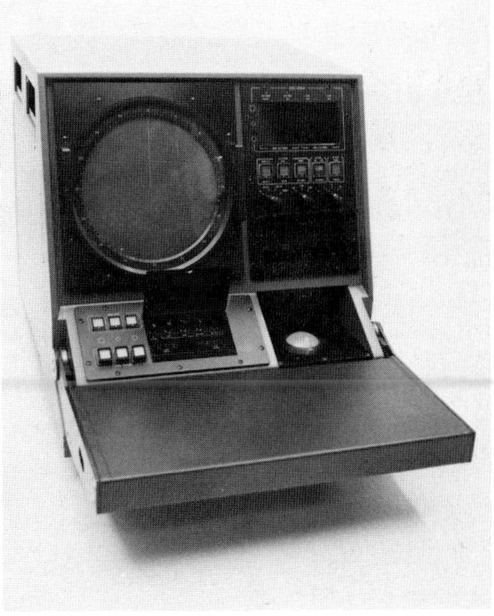

Operator's console for the LPD/20 radar.

The LPD/20 has been designed to meet the following requirements: use of the fully coherent pulse doppler technique to eliminate fixed echoes and to achieve a very high MTI improvement factor. High information rate to reduce reaction time of the associated anti-aircraft weapon systems and to facilitate the engagement of multiple targets in rapid succession. X-Band working frequency to achieve a narrow beamwidth with the same antenna size whilst ensuring a high angular resolution and high accuracy in target designation for azimuth. Appro-

priate pulse length for a small range gate, necessary for accurate target designation in range. Target detection range of up to 20km. The LPD/20 can be integrated with IFF equipment. Maintenance — built in test and calibration units, use of modular plug-in units and rapid test facilities. Reliability is enhanced by the use of solid-state components, except for the power transmitting tube.

The LPD/20 search radar operational.

Employment

By 1979 Contraves Italiano had built in excess of 100 LPD/20 units.

Minilaser Rangefinder

Netherlands

In addition to its well-known range of laser rangefinders for artillery and tanks, Oldelft (nv Optische Industrie De Oude Delft) has recently developed a lightweight Minilaser rangefinder for field use by forward artillery observers. New for this type of instrument is the digital readout of distance to the target and azimuth and elevation of the illuminated indications which allow the instrument to be used in twilight. The Minilaser is small and very light and a quick-release fastener enables the unit to be quickly mounted on the tripod for use in the field or on the optional vehicle adapter when the rangefinder is to be used on a tank or Jeep. The vehicle adapter itself has a universal design and can be attached simply to any flat surface.

The Minilaser has been designed to measure distances up to 10,000m with a rate of 10pulses/min and has a first and last reply mode. The main components are the rangefinder and the goniometer which are housed separately and can be used separately. The 12V NC battery has sufficient power for at least 250 sightings. The observing telescope has a magnification of ×7 and a field of view of 7°.

The Minilaser rangefinder in operational use.

The weight of the rangefinder with goniometer is 7.8kg, the battery weighs 2.4kg and the tripod 3.3kg total weight of the complete equipment is approximately 13kg.

L4/5 Weapon Control System — Netherlands

Weight: 6,000kg
Length: 7.5m
Width: 2.2m
Height: 5.2m (operational)
7.5m (travelling)
Range: 31km (max search radar)
34km (max PPI)
32km (max tracking antenna)
Target speed: 500m/sec (max horizontal)
300m/sec (max vertical)
Operating conditions: temperture from −25° to +110°F
Winds up to 75km/h
Crew: 2/3

The L4/5 all-weather fire control system.

This system was developed and built by N. V. Hollandse Signaalapparaten of Hengelo in the late 1950s and early 1960s. Extensive trials were carried out in 1962 and 1963 and these proved that the system was an effective answer to low level attack, it can be used with both anti-aircraft guns (40mm) or guided missiles. Normally one L4/5 controls three 40mm guns. The complete system is all on one trailer, although another trailer is used for the generator. The six main components of the system are the search antennas, tracking antenna, monocular sight, operators' cabin, velocity measuring equipment and the converter.

There are two search antennae back to back, one of these is for medium and low targets and the other for very low targets, one tracking antenna and a monocular sight for optical interception and location. The control panel consists of the panel with a PPI for medium and low targets and a PPI for very low targets, an A-scope for hand tracking, testing and operational controls and the control panel. The digital transistorized computer is under the cubicle.

Its main features are the detection of aircraft at very low to medium heights, extensive anti-jamming features, IFF facilities, short reaction time, rapid search scan, automatic picking-up and tracking of targets, search-while-track through 360°, digital computer programmed for the simultaneous control of three guns, possibility of automatic connection with a central position by means of digital data links, simple operation, high degree of mobility.

FASCAN Search Pattern. The system is so designed that all targets within the search range are detected at an adequate range. The PPIs provide a clear survey of the tactical situation. The spiral scan pattern of the high beam can be tilted through about 3.4°. The axis of the low beam can be tilted between −0.8° and +3.4°. The beams thus overlap.

Modes of Operation — the search radars can be either continuous search or sector search. Picking-up, automatically by radar or manual, ie the director operator can pick up any target visually with the aid of two handles with built-in grip switches. Tracking — automatic or manual. In automatic it can either auto-radar, generative or free generation. In manual it can be optical or on the PPI.

Employment

Production complete, in service only in the Netherlands.

Flycatcher All-Weather Fire Control System — Netherlands

The Flycatcher has been developed by Signaal (Hollandse Signaalapparaten BV) to fulfil a requirement for an all-weather fire control system that could be used with both guns or surface-to-air missiles. The complete system is mounted in a fire-resistant container which can be transported by transport aircraft or helicopters. For transport the antennas and TV camera are electrically retracted inside of the container and the petrol driven generator can be carried inside the container.

The radar of the Flycatcher consists of an integrated search and tracking unit, operating on the X band and providing search-whilst-track capability. This radar is supplemented with a separate anti-image tracking radar which permits accurate and continuous tracking at near ground elevation angles. Switching between these radars occurs fully automatically. For search, a coherent MTI receiver is used consisting of a combination of an MTI oscillator, a coherent oscillator (COHO) and a double digital canceller. This is followed by a video correlator and a pulse-length discriminator.

The display and control group comprises a PPI on which the moving targets and certain markers are displayed and a scope which gives the operator tracking information from both radars together with three rows of numerical indicators presenting on request target data, tracking values, muzzle velocity or gun parallaxes. This group also contains the joystick for target designation and the operational push-buttons. An input data unit, equipped with decimal switches, transfers semi-variable data such as parallaxes and meteorological information to the computer. The display ground can also have a TV monitor. The computer group consists of a mini-computer SMR-S, an interface and a power supply unit.

Employment

In service with the Dutch Air Force.

The Flycatcher all-weather fire control system.

Weight: 2,700kg
Length: 2.73m
Width: 2.12m
Height: 2.13m (antenna in)
3.65m (antenna out)
Reaction time: average 4.3sec
Search coverage: up to 20,000m, 360°
PPI scales: 10 and 20km
ECCM: simultaneous with clutter rejection
Operating temperatures: –40° to +52°
Search antenna speed: 40rpm

Reporter Search Radar

Netherlands

The Reporter search radar has been developed as a private venture by Hollandse Signaalaparaten to meet a requirement for a search radar which can provide light anti-aircraft guns with information as to speed and direction of approaching hostile aircraft. The radar, which operates in the 3cm band, is trailer mounted and the antenna can be lowered for transport. The complete system can be towed by a LWB Land Rover or similar vehicle which also contains the radar screen and operating controls. The prototype of the Reporter is currently undergoing manufacturers trials.

Employment

Trials.

Artist's impression of the Reporter search radar deployed.

Simrad LP3 Laser Rangefinder

Norway

The Simrad LP3 laser rangefinder is a one-man portable, electro-optical instrument designed for use by artillery observers. It contains in one unit the laser transmitter, receiver system, electronic circuits, sighting telescope and goniometer. Total weight is 6.4kg.

The instrument instantly measures range, azimuth and elevation angles to possible targets and may be used for determination of opposition, target position and for adjustment of fire. It can measure ranges from 200-20,000m with an accuracy of ±5m. The azimuth angle range is 6,400mils and elevation angle is ±350mils, both with a resolution of 1mil.

The rangefinder fits on a special lightweight tripod, designed for field and combat conditions. The tripod can be adjusted heightwise, and the range finder fits to it by means of a tripod screw.

The battery pack consists of NiCd cells and is clamped to the tripod, the battery has a rated voltage of 24V and its capacity is some 600 firings at +20°C. The battery is rechargeable.

A special sighting telescope with reticle is combined with the optical receiver. Under normal conditions the rangefinder is able to range any target within 6,000m that will produce an image covering an open one mile wide area in the centre of the reticle. The reticle is illuminated by beta light for night use.

The 20° angle of the eyepiece allows the operator to wear a helmet when he aims the instrument through the sighting telescope. The top of the range finder housing has a quick-aiming sight which enables the range finder to be rapidly pointed against targets appearing only for a short time. The range finder housing is waterproof.

The operation of the instrument is based upon the echo principle. When the trigger is activated, a short, invisible light pulse of high intensity is transmitted from the laser optics to the exit window. When hitting a target, some of the transmitted light is reflected back towards the instrument, it passes through the entrance window and activates the receiver.

The range is displayed directly in metres and up to three individual targets (if located at least 30m apart) can be displayed simultaneously. The display also provides indication of targets within the minimum range, low battery voltage and low laser power. The display will light for 10sec, then extinguish.

The minimum range sets the range within which reflections from targets shall be rejected. The minimum range may be set at any distance between 200 and 6,000m. Total weight of the instrument including tripod is about 11kg.

The Simrad laser rangefinder LP3.

The LP3 has been developed by Simrad AS of Norway for the Norwegian Army under a contract with the Army Material Command. The development was supported by the Norwegian Defence Research Establishment. The Instrument is now in series production for both the Norwegian and British Army (as the LP6) as well as for armies in other countries.

Simrad LP7 Laser Rangefinder

Norway

The Simrad LP7 laser rangefinder has been developed to enable infantry units accurately to range targets for engagement by close support weapons such as mortars. The LP7 entered production in 1978 and has already been ordered by the British and Norwegian armies.

The instrument weighs only 1.7kg and range is determined by laying the reticle of the aiming telescope on the target and pressing the fire button on top of the unit. Range is immediately displayed in the eyepiece and unwanted echoes may be gated out from the display by means of a minimum range control located on the operators side of the instrument. A small built-in rechargeable 12V battery provides 600 rangings on one battery charge.

The transmitter is a miniaturised Q-switched Nd : YAG laser. The sighting telescope is combined with the optical receiver by use of a special beam

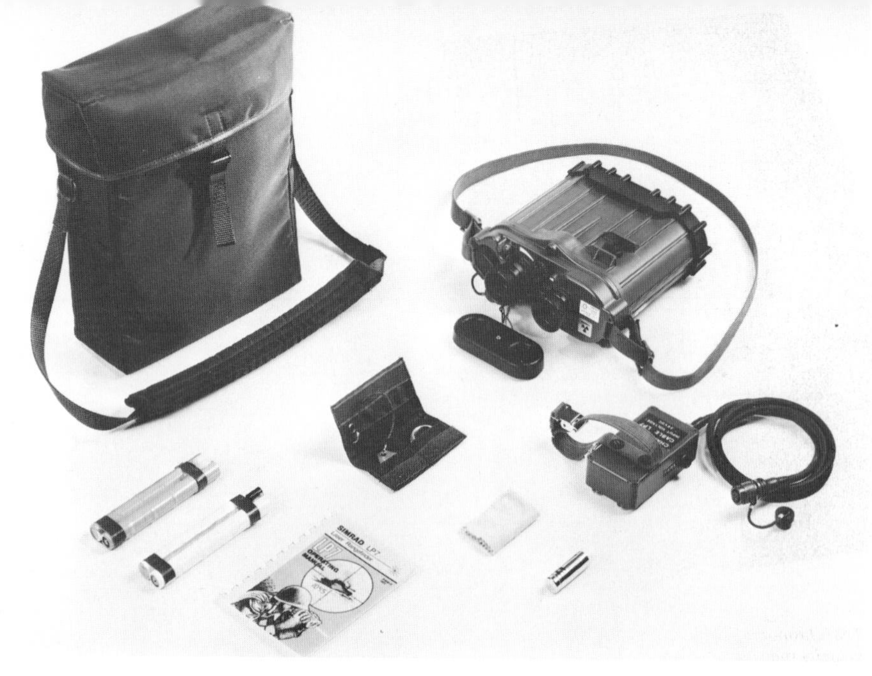

splitting technique. The performance of the sighting telescope is comparable to that of a standard observation monocular. The receiver uses a Silicon Avalanche-photodiode and this gives the instrument a range capability of up to 9,000m. To reduce the safe distance of the rangefinder, the output energy from the laser transmitter is kept to a minimum.

The four digit LED display is observed through the left eyepiece and is superimposed on the picture given in the other eyepiece. The intensity of the display can be varied by rotating the eyepiece housing. The display is switched on when a ranging has been performed and is automatically switched off after 3sec.

The Simrad laser rangefinder LP7 and its accessories.

NM87 Muzzle Velocity Radar

Norway

The NM87 muzzle velocity radar has been developed by the Norwegian Defence Research Establishment under contract to the Norwegian Army Material Command and is manufactured by Nera Bergen.

The system consists of four main components: Doppler radar (weight 13kg inc mounting bracket), chronograph (3.6kg), cable reel (12kg) and the mounting set. The doppler radar unit houses the transmitter receiver and is mounted directly on the gun carriage. The chronograph unit incorporates the logic units, solid state numeric display, the power distribution circuits and all operational controls. The cable reel contains two cables each 30m long. One power cable for battery connection and one cable for power and signal distributions to both the Doppler radar and the chronograph. The mounting set consists of a bracket with the necessary supports and screws to mount the Doppler radar to the gun.

The Doppler radar transmits continuous power at a wavelength of approximately 3cm and is mounted by means of a bracket on to the gun carriage and so follows the gun in elevation. A cable connects the transmitter/receiver unit with the indicator unit (the chronograph) from which all operations are performed, a separate cable connects the

The Doppler radar mounted over the barrel of 155mm M114 towed howitzer.

The chronograph set used with the NM87 muzzle velocity measuring equipment.

chronograph with the 24V power supply. By means of a parabolic antenna a radar beam is transmitted along the trajectory of the projectile. When the projectile leaves the muzzle of the gun and enters the radar beam some of the transmitted power is reflected and is detected in the receiver. Because of the velocity of the projectile the reflected power has a lower frequency which is filtered out in the receiver. By counting the Doppler periods the position of the projectile is determined completely independently of the velocity.

After 2,176 Doppler periods, corresponding to 35m, an electronic gate is opened for a duration of 128 Doppler periods, corresponding to a base length of about 2m. During this time a decimal counter counts the clock pulses from a quartz-controlled clock. Thereby the flying time of the projectile is measure in units of 0.25 micro-seconds. This time is indicated on a five-digit solid state numeric display. The velocity in m/sec (150-850m/sec) is then found by means of a table which has been calculated from the Doppler equation. The velocity at the muzzle is then found by another table which is based upon ballistic calculations under standard conditions.

Employment

Norway.

Odin Fire Control System

Norway

ODIN is the name given to the artillery fire control system at present used by the Norwegian Army. The complete system comprises a laser rangefinder, manufactured by Simrad A/S (see separate entry), a meteorological system (including weather radar), a muzzle velocity measuring system, developed by Elektrisk Bureau A/S, Division Nera, an electronic transmission device, developed by Siemens A/S, Norway, a gun display and a SN-302M General Purpose Military Computer manufactured by Kongsberg Våpenfabrikk. The latter consists of a central processing unit, a memory unit, a field artillery panel and a power supply unit. This weighs only 70kg and requires a 24V power supply. It is normally mounted in a truck or a M577 tracked command vehicle.

The SN-302M is deployed both at battery and battalion headquarters level. It comprises the

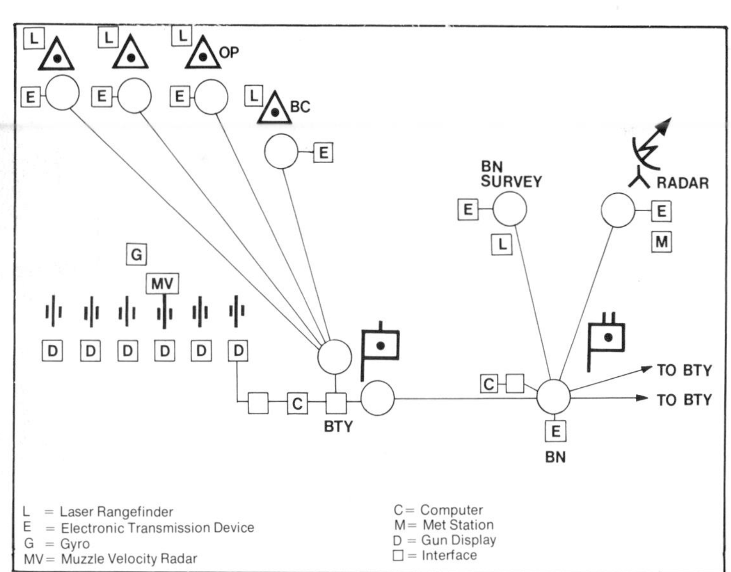

The complete Odin artillery fire control system.

following: master controls (start button, tapeload button, programme category selector switch, programme name display and so on), the matrix, data display and the keyboard itself. The programmes are categorised into the following related groups — topographic data entry, survey computation, gun and ammunition data entry, meteorological data entry, firing data computation and storing data.

The SN-302M computer in action.

Ericsson Hand-Held Laser Rangefinder — Sweden

Transmitter type: Neodymium-YAG with passive Q-switch
Wavelength: 1.06μm
Pulse length: 12nsec
Pulse repetition frequency: 0.5Hz
Range capability: 5,000-6,000m against ground targets
Range accuracy: ±5m
Extinction value: 30dB
Sight: 100mrad (field of view) ×6.5° (magnification)
Weight: 2.5kg

The Ericsson handheld laser rangefinder.

The Ericsson hand-held laser rangefinder has been designed not only for use by forward artillery observers, but for any application where there is a need for fast and accurate range determination. It is similar in appearance to ordinary field glasses and uses a high-quality optical sight.

The housing contains the small laser sub-units developed by Ericsson and the rechargeable and easily replaceable battery. When the firing button is depressed the laser instantly displays the target value in the eye-piece. The laser has two counters which makes it possible to range two targets simultaneously. An adjustable minimum range is provided to assist the operator when selecting between two different target ranges. If required, the hand-held laser rangefinder can be mounted on a tripod with a goniometer by means of a mechanical mounting. For operations under low-light conditions the graticule is illuminated.

Employment
Prototype.

Ericsson Artillery Laser Rangefinder

Sweden

Transmitter type: Neodymium-YAG with passive Q-switch
Wavelength: 1.06μm
Pulse length: 12nsec
Pulse repetition frequency: 0.5Hz
Range capability: 10,000m against ground targets 6,000m against shell bursts
Range accuracy: $\pm$5m
Extinction value: 40/dB
Sight: 130mrad (field of view) x8 (magnification)
Weight: 11kg

The Ericsson artillery laser rangefiner.

The Ericsson artillery laser rangefinder has been designed for use by field artillery forward observers. The equipment consists of a laser transceiver, goniometer for the accurate determination of azimuth and elevation angles, a tripod and a battery unit. The tripod legs are quickly adjusted and levelled to enable the observer to keep a low profile. The adjustable azimuth scale is set to a known direction and the laser is ready for operation.

The transceiver consists of new generation laser sub-units developed by Ericsson. The laser trigger button is placed on the left side of the goniometer, under a protective cover, to prevent accidental firing. The target range value is presented in the left hand side eye-piece and the azimuth and elevation scales are placed on the goniometer. A continuously adjustable minimum range is provided to assist the operator when selecting the correct target range. For operation under low-light conditions the graticule and the angular scales are illuminated.

Employment
In production.

SAAB ACE-380 Artillery Computer Equipment

Sweden

The ACE-380 is a Field Artillery Computer Equipment for firing data calculations and has been developed by Saab-Scania. The system can be integrated with existing fire control systems. It operates at battery level but can also supply firing data to neighbouring batteries.

Its main characteristics are: single unit design, portable by two men, simple one-man operation, general purpose digital computer (Datasaab D5-30), built in fault detection facilities, fixed or moving target data calculation, firing data within one second and easy training.

The use of a multi-purpose digital computer with ferrite core memory and real-time programming form the basis for a number of tactical and operational capabilities, from the point of view of battery staff nothing is new, only much faster.

Its capabilities include target data storage for 500 targets, firing data for up to 40 ballistics inclusive of

The SAAB ACE-380 in operational use.

corrections and daily adjustments, storing of complete target data with corrected firing information for 20 targets, automatic or semi-automatic input from forward observer, automatic output of firing data to gun data link, automatic input from transmitter of muzzle velocity data, prognosis of air for 30 levels above the ground. All data is displayed immediately when required and is easily updated.

The keyboard and the display panel are designed with separate functional fields for each group of parameter data input. All input data can be checked on the panel before executing orders are released.

The keyboard is divided up into the following operational fields: fire adjustment, changing of target, battery position data, target data, ballistic data and influences (ie air pressure, muzzle velocity, projectile weight etc). There are also clear, store compute, test, fire etc buttons. In addition there is an alpha numerical key area.

Employment

The ACE-380 has been tested by the Swedish and one foreign Army but has not so far been placed in production.

Model 9 FA301 Field Artillery Computer — Sweden

The Field Artillery Computer Model 9 FA301 has been developed by Philips Elektronikindustrier AB and is a further development of the earlier 9 FA101 Field Artillery Computer whcih has been in service with the Swedish Army for some years.

The main function of the 9 FA301 is to compute firing data including bearing, elevation, fuze setting and charge. This data is presented on the LED displays on the display panel and is also transmitted to the RIA gunsights (qv) or to repeater displays at the guns. Data transmission can take place within two seconds from the input of the complete data.

The 9 FA301 is built as one unit in modular design comprising a general purpose digital computer with micro-programmed CPU, input/output units, 32 k words core memory and ergonomically designed control/display panels. The complete unit can be mounted in a vehicle or can be set up separately. Only one man is needed to operate this equipment. A paper tape reader can for maintenance purposes be used to load programs into the core memory. An alpha-numeric printer can also be supplied with computer as an optional extra. This can be used for supervising the firing procedure and in addition is useful for carrying out effective training of the operator.

Main operational functions of the Model 9 FA301 which weighs 45kg, can be summarised as battery and gun position registration/deregistration, meteorological and ammunition data registration, target data registration/deregistration, firing data computation for fixed and moving targets, computation of individual gun corrections, firing limits registration, ammunition follow-up registration, continuous testing of program and automatic/manual printout of entered and computed data.

The field artillery computer Model 9 FA301.

Employment

In service with the Swedish Army.

Model 9 FA101 Field Artillery Computer — Sweden

This is a digital fire control computer for use within the field artillery. The system calculates firing data for guns for use against both moving or stationary targets. The system has been developed by Philips Elektronikindustrie AB, development started in 1966. The first production contract was awarded to Philips in 1971 and the system entered service with the Swedish Army in 1973/74. It weighs only 45kg and is deployed at battery level. The main functions comprise — calculation of firing data on the basis of manually set or automatically fed input values, tactical switching between a maximum of 28 selectable ballistic inputs, input of fire observation data in accordance with the line-of-collimation principle, computation of coefficients for the utilisation of fire correction, input of altitude observation data for fuze setting and the input of speed and course of moving targets.

The main settings on the 9 FA101 are ballistics, gun and target coordinates, ballistic influences, fire and trajectory corrections and for moving targets speed, course and shooting interval. The computed firing elements (bearing, elevation and fuze setting) are presented on the indicator tubes of the panel.

Philips have also developed the Gun Indicator. This is mounted on the gun in front of the gunner and consists of a data receiver and a number of registers for the storing and presentation of firing data. Firing signals, including those for firing at moving targets are given by lamp and buzzer. Another lamp indicates when new firing data has been computed. Firing data and voice communication are transmitted simultaneously on the same two-wire circuit.

Finally Philips have developed the HADAR Data Transmission system which transmits information between the forward observer and the 9 FA101 computer.

Employment

In service with the Swedish Army.

The Philips field artillery computer Model 9 FA101.

RIA Automatic Electronic Gunsight

Sweden

Weight: 41kg (sighting unit)
8kg (panoramic sight)
21kg (direct sight)
25kg (control and data unit)
Magnification × 4 (panoramic and direct)
Field of view: 7.5° (panoramic)
10° (direct)

The RIA Automatic Electronic Gunsight has been designed and developed by Philips Elektronikindustrier AB for use with the Bofors 155mm FH-77 weapon, but can, in modified version, be mounted on most modern field guns and howitzers.

The RIA comprises a control and display unit and sighting unit with a panoramic and a direct sight for

The RIA automatic electronic gunsight with display unit on the right.

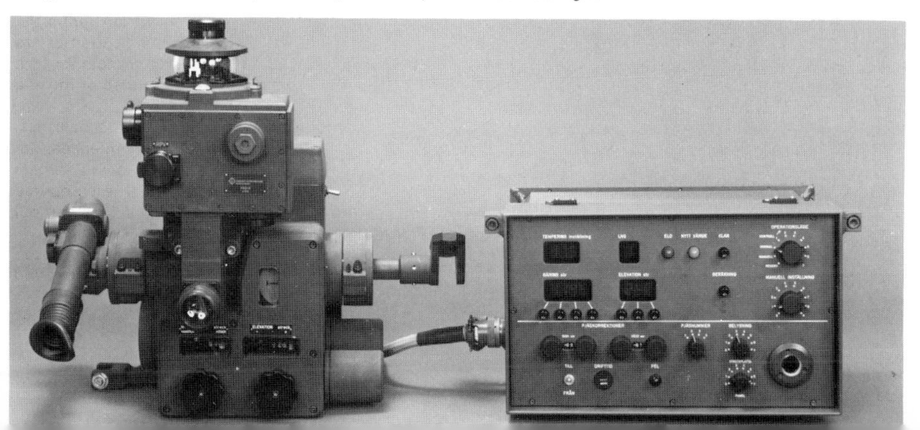

use in the indirect and direct fire roles. The main operational functions of the RIA can be summarised as the presentation of firing and gun laying data, automatic control of line of sight in azimuth including correction for tilt of the gun, manual settings of individual gun corrections in azimuth and elevation, reserve mode for use when the data transmission is out of action and reserve mode for use when the electronics are out of action.

Employment
In service with the Swedish Army.

Ericsson Giraffe Search Radar — Sweden

Giraffe is the family designation of a mobile search pulse radar system developed by L. M. Ericsson. The basic system, called the PS-70R, has been designed for use with the Swedish Army Bofors RBS-70 short range surface-to-air missile system, it can also be used in conjunction with light anti-aircraft guns.

The complete radar, together with radio communications equipment, tactical control facility and power supply, is linked with a number of firing units for precision target designation and combat control.

The equipment is mounted in a self-contained container which is carried on the rear of a cross country vehicle, in the Swedish Army the Saab-Scania SBA 111 (4×4) or SBAT 111S (6×6) truck is used.

A feature of the Giraffe is the use of the folding mast for the antenna giving an operating height of 12m. This together with excellent clutter suppression characteristics, gives the Giraffe very low altitude coverage. The range coverage is 40km in the surveillance mode and 20km in the target designation mode. The radar is offered with comprehensive ECM options and IFF equipment can be installed .

The radar operates in the C-band and digital doppler processing and constant false alarm circuitry are used to automatically extract, detect, and present the target of interest. A digital plan position indicator presents video signals and target tracking signals. The target data link constitutes the interface between the search radar and the firing sites. The latter can be located up to 5km from the search radar. Information about target position, speed and course is derived and transmitted from the target data transmitter which is located in the search radar. The target data receiver, located at the firing site, performs certain calculations and generates signals for slewing the gun or missile sight towards the target and displays information as to whether the target is engageable or not.

Employment
The PS-70/R is in service with the Swedish Army while the Giraffe is in service with a number of overseas countries including at least two in Africa.

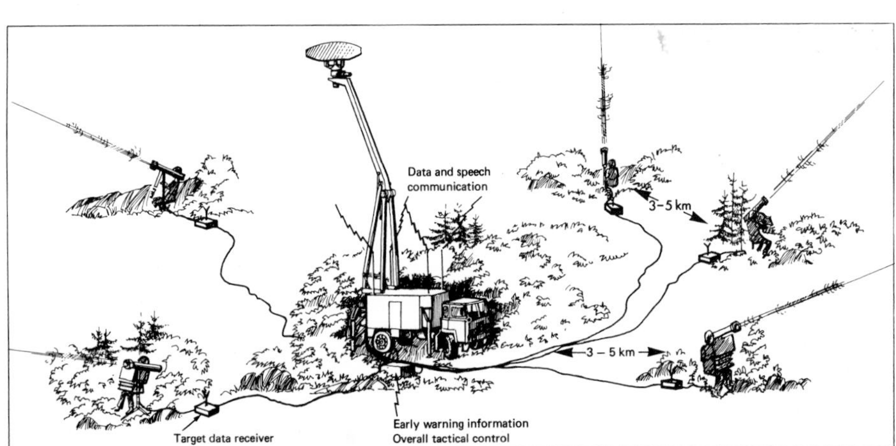

The PS-70R being used to control five Bofors RBS-70 SAMs.

Skyguard AA Fire Control System — Switzerland

Skyguard has been developed by Contraves AG in co-operation with L. M. Ericsson of Sweden and Albiswerke Zurich AG. It is a mobile miniaturised all-weather fire control system for use against low-flying aircraft and surface to air missiles. It can be used to control anti-aircraft guns or guided missiles.

The system consists of the following:

Pulse Doppler Search Radar. This has high detection probability, high information rate, fine resolution cell, high clutter suppression, automatic target alarm, automated acquisition procedure for tracking radar, resistant to ECM, elaborate video

Skyguard ready for operation

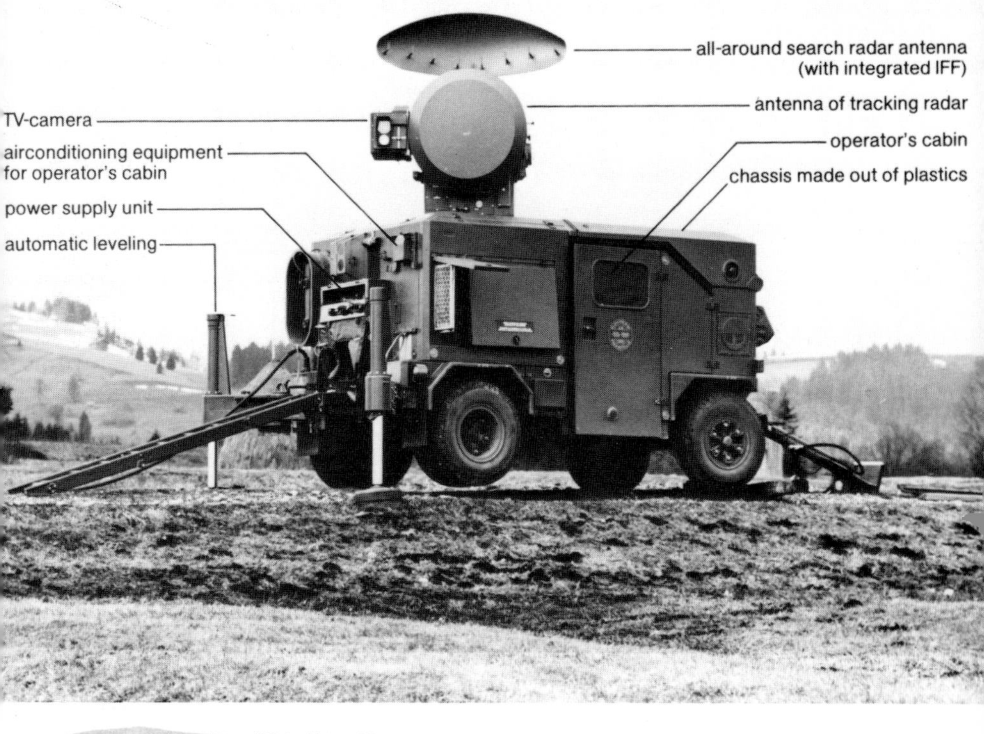

Skyguard in operation showing main components of the system.

presentation and can have IFF facilities. Max range of the search radar is 20km.

Pulse Doppler Tracking Radar. Monopulse-evaluation, fine resolution, accurate tracking, automated target exchange under computer control, ASM detection and alarm, resistant to ECM and second tracking system when tracking missiles.

TV Tracking System. Automatic high-precision target tracking, using video-processor, computer aided tracking using a joystick Vidicon (with or without light intensifier) or SEC-Vidicon camera available. The tracking mount supports the tracking antenna and TV Camera as well as the independently rotating search radar.

Contraves Cora 11 M Digital Computer. This carries out the following tasks: threat evaluation based on search radar information, hitting point data for the guns, calculations for guided missile deployment, calculations of tactical target data, monitoring of all Skyguard sub-systems. Full automatic functional and performance check of the entire system. Simulation of combat situations for realistic training.

Digital-Data Transmission. Standard two-wire field telephone connection cable to the weapons.

Control Console. This can be operated by one or two men and has an indicator (PPI) displaying doppler and raw-video as well as symbols and markers for the presentation of the tactical situation. A scope is provided to check target tracking and to judge the ECM situation. Tactical display with numerical read-out, monitoring of the TV tracking system, rolling-ball to control markers, joystick for manual control of the tracker, matrix panel for data input and output, and an inter-communication system.

Power Supply System. This is mounted internally but can be removed for external operation, it consists of an air cooled, four stroke petrol engine

The system is mounted in a body made of fire-resistant reinforced fibre-glass polyester and is fully air-conditioned. The tracker folds away whilst travelling. It has automatic hydraulic fine levelling and has good accessibility for maintenance purposes. The body is air-transportable and can be mounted in the rear of various types of vehicle.

Employment Skyguard has been ordered by seven countries including Austria, Germany (FGR), Spain and Switzerland, and is now in production.

Super-Fledermaus AA Fire Control System — Switzerland

Done

Weight: 5,000kg (approx, trailer)
Length: 4.6m
Width: 2.3m
Height: 2.75m
Tracking speeds: 2,000mil/sec (in bearing)
1,000 mil/sec (in elevation)
Range: 200m (min measuring)
50km (max scanning)
40km (max tracking)

The Super-Fledermaus Anti-Aircraft Fire Control System has been developed by Contraves AG in association with Machine Tool Works Oerlikon Bührle (who developed the muzzle velocity measuring equipment), and Albiswerk Zurich AG.

The complete equipment is mounted on a four-wheeled cross-country trailer and consists of the following:

(a) Tracker — with one man control, telescopic sight and radar antenna for target acquisition and tracking.

(b) Electronic Computer — continuously determines the firing data for three individual gun emplacements.

(c) Acquisition and Fire Control Radar — with turntable magnetron transmitters which can be switched over during action in case of jamming.

(d) Muzzle Velocity Measuring Equipment — measures the muzzle velocity of the three guns individually.

(e) Optical Putter-On — This enables the tracker to be directed immediately on to targets appearing unexpectedly.

(f) Signal box — enables the fire control officer to give orders for target acquisition, tracking and opening fire by remote control.

The system can acquire targets up to 50km away by night or day and track them automatically from a range of 40km. It can do this by the following five modes of operation:
(a) Helical scanning, the antenna rotates in bearing and slowly threads its way down. In this manner the entire horizon is scanned in a gapless helical sweep. (b) Sector scanning, a vertical oscillating motion is imparted to the antenna, at the same time the antenna is slowly rotated in bearing so that the radar beam scans a specified sector of space without a gap by rowing one path to the next. (c) Early warning radar, the fire control unit can be put on in bearing and in range by an early-warning radar station. (d) Visual Target Acquisition, independent picking up of a target by the optical layer who controls the tracker by means of a joystick and a telescope. The range is adjusted by the radar operator. (e) Target acquisition with external optical putter-on (see above). As soon as the target has been detected the radar beam is locked on to the target and the unit can be switched over to automatic tracking. There are four methods of tracking a target by this system: (a) Automatically by radar (b) Optically by visual layer. (c) Automatically without radar, with regenerative control. (d) Automatically without radar but with memory.

Super-Fledermaus deployed for action.

The radar operates as a micro-wave pulse radar and was designed by Albiswerk, Zurich AG. The search indicator can be switched over automatically or manually to Plan Position Indicator or Range Height Indicator as desired. The range indicator which shows on two simultaneously visible traces both the entire range from 0-40km (A-display) and a greatly expanded zone of ± 1km around the target (R-display).

The Super-Fledermaus analogue computer determines the firing data for the guns. From the moment a target is tracked by the fire control radar of the visual tracker, the present position data are continuously fed into the computer, this determines the data and transmits the data to the guns through cables. The computer also allows for the following: individual horizontal and vertical parallaxes for three different gun emplacements, wind speed and wind direction, air density correction and individual muzzle velocities for each gun. Test and calibration instruments are built into the system, this allows a thorough control of the important electrical functions at any time.

Employment

Super-Fledermaus is used by more than 20 countries including Argentina, Germany (FGR), India, Italy, Japan and Switzerland, over 1,500 units have been built to date. It is built under licence in India, Italy and Japan and can be used in conjunction with 35mm or 40mm AA guns.

Fieldguard Fire Control System

Switzerland

Fieldguard fire control system deployed for action; the vehicle used in this case is a MAN (4×4) truck.

The Fieldguard (previously known as the CONAR) Fire Control System has been developed to prototype stage by Contraves AG and Siemens-Albis AG of Switzerland.

The system calculates the co-ordinates associated with the gun or multiple rocket system (eg the German LARS) battery and determines the firing data without weather information by employing pilot shots. Evaluation, azimuth and fuze setting can be calculated individually for up to eight guns or launchers. Basically one of the guns or launchers fires a pilot round and this is followed by the Fieldguard's radar for a part of its trajectory (between two-thirds and three-quarters). After this pilot round is destroyed in mid-air and the point of impact is calculated by the computer.

This calculation allows the CORA II computer to calculate correction values for the real target which are transmitted direct to the weapons by data transmission units. One of these is with each gun and consists of a four digit display giving aiming data including azimuth, elevation and fuze setting.

Components of this system include the Fieldguard operator's console which has the following main parts — control radar panel, display and operator panel radar, pilot and tactical display panel, data input keyboard and the display panel computer.

The precision radar is mounted on the roof of the cabin and the computer installed is the Cora II. The system has a built in power supply which comprises a four stroke petrol engine.

The complete Fieldguard Fire Control System is mounted in a fully air-conditioned cabin which can be mounted on a standard army truck, fitted to a M548 tracked supply vehicle or mounted on a trailer. It can also be transported by helicopter.

Employment

Fieldguard has been evaluated by the German (FGR) Army which is expected to place a production order for the system in the near future.

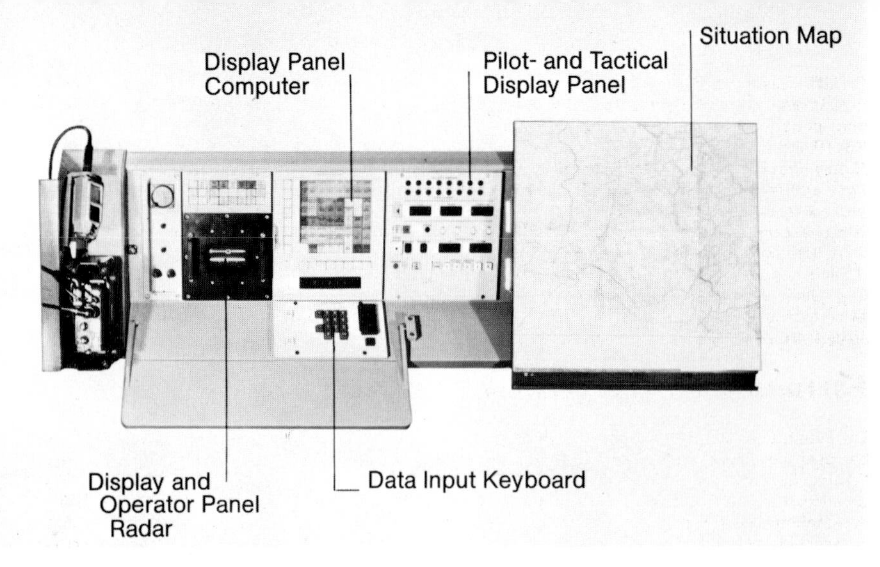

Operator's control console of the Fieldguard system.

Artillery Fire Control Equipment USSR

DIRECTORS

PUAZO-4 Introduced in 1945 and used with the 85mm M1944 anti-aircraft gun.

PUAZO-5 Used with 57mm S-60 anti-aircraft gun and SON-9/SON-9A ('Fire Can') radar, but subsequently replaced by PUAZO-6/60 director.

PUAZO-6/12 Used with 85mm M1939 anti-aircraft gun and SON-9 ('Fire Can') radar, replaced the earlier PUAZO-3 director.

PUAZO-6/19 Used with 100mm KS-19 anti-aircraft gun and SON-9 ('Fire Can') radar

PUAZO-6/60 Used with the 57mm S-60 anti-aircraft gun and SON-9 ('Fire Can'), replaced earlier PUAZO-5 director.

PUAZO-7 Used with 100mm KS-19 anti-aircraft gun and SON-4 ('Whiff') radar, replaced by PUAZO-6/19.

PUAZO-30 Used with 130mm KS-30 anti-aircraft gun and SON-30 radar.

RADARS

'Fire Can' This trailer mounted system was first deployed in 1954 and is known as the SON-9 (or SON-9A) by the Soviets. It has been used in conjunction with the 57mm S-60 anti-aircraft gun (with PUAZO-6/60 director), 85mm M1939 anti-aircraft gun (with PUAZO-6/12 director) and the 100mm anti-aircraft gun KS-19 (with PUAZO-6/19 director). It was also used with the PUAZO-5 director and 57mm S-60 anti-aircraft gun.

'Fire Wheel' This trailer mounted system operates in the I-J bands and is known as the SON-30 by the Soviets. It is used in conjunction with the 130mm KS-30 anti-aircraft gun and the PUAZO-30 director.

'Pork Trough' radar mounted on AT-L light tracked artillery tractor.

'Whiff' This trailer mounted system operates in the E-band and is known as the SON-4 and is used in conjunction with the 100mm KS-19 anti-aircraft gun KS-19 and the PUAZO-7 director.
'Long Trough' This E-band radar is mounted on a light artillery tractor and is known as the SNAR-1 artillery reconnaissance radar by the Soviets.
'Pork Trough' This J-band radar is mounted on an AT-L light tracked artillery tractor and operates in the J-band, maximum range is believed to be 9,000m. The Soviets call the 'Pork Trough' the SNAR-2 artillery reconnaissance radar.
'Track Dish' This equipment is mounted on an AT-T heavy tracked artillery tractor and is known to the Soviets as the ARSOM-1 artillery and mortar position locating radar.
'Small Yawn' This equipment is mounted on an AT-L light artillery tractor and operates in the I-band. The Soviets call the system the ARSOM-2 artillery and mortar locating radar.
'End Tray' This equipment is mounted on a four wheel van type trailer and operates in the D-band. It is known by the Soviets as the RMS-1 meteorological radar and is used in connection with surface-to-surface missiles (such as the 'Scud') and well as conventional artillery.

Ferranti Laser Target Marker and Ranger — UK

The Ferranti Laser Target Marker and Ranger (LTMR) has been developed by the Laser Systems Group of Ferranti Limited, under contract to the Ministry of Defence. The LTMR is the ground part of the laser-designating close of air support system and is now in service with the British Army.

Basically the Forward Air Controller (FAC) aims the LTMR at the target and switches on as the aircraft approaches. The airborne-seeker acquires and tracks the laser energy scattered by the target. In the case of aircraft using conventional weapons, the pilot's head-up-display is controlled to indicate target position and the weapon may then be released. When laser guided weapons are used the bomb or missile homes on the energy scattered by the target.

The FAC has the following equipment — the LTMR, Precision Angulation and Support System, battery power supply and carrying harness. It has a Neodymium liquid cooled laser and has an accuracy of ± 5m out to 10,000m, target marking for at least 20min before the battery runs out (an average mission however would last less than 60sec). Three ranges can be displayed simultaneously. The basic LTMR weighs only 7.25kg and the battery a further 3.2kg.

The system can also be used in conjunction with a Night Observation Device developed by Pilkington P. E. Limited.

Although the LTMR has been designed for the air-to-ground role, it can also be used as an artillery laser range finder and for directing the Copperhead CLGP.

The Ferranti Laser Target Marker and Ranger.

Ferranti PACER Muzzle Velocity Measuring Equipment — UK

The PACER has been developed by the Cheadle Heath Division of Ferranti Computer Systems Ltd and has been designed specifically for use by forces in the field. After setting up the equipment all the operator is required to do is to set one control for the approximate anticipated muzzle velocity and press a 'reset' button. The operation of the equipment is then automatically initiated by detection of the instant of fire by the muzzle flash detector. Within two seconds of firing, the achieved instrumental muzzle velocity is presented on a legible decimal display in metres/second units. The radar transmitter is activated at the instant of fire and is switched off automatically when the velocity measurement is completed. This is useful when enemy sensors are in the area.

The equipment incorporates circuitry which automatically checks the conditions under which each measurement is made are satisfactory. If these checks indicate that the measurement may be invalid an all-zero result is indicated. It also incorporates an overall test facility. The equipment will operate from a 24V vehicle supply and will operate in

PACER. On the left is the aerial and on the right the display unit ready for use. Cables are not connected in the photograph.

temperatures −20° to +55°C.

PACER is accurate to ±0.30m/sec for velocities between 200 and 1,599m/s provided that:

(a) the anticipated velocity control is set within ±20m/sec of the true velocity.

(b) The aerial is sited correctly 2m ±0.5m from the barrel axis between breech and muzzle and aimed along the line of fire with an error less than ±4° in elevation and ±8° in bearing. The built in clinometer gives adequate elevation accuracy.

The four components of PACER are: (a) Flash detector, weight 1.50kg (b) Aerial, weight 40kg (c) Display unit, weight 34kg (d) Cable and drum, weight 45kg.

Employment

The PACER is fully developed and is in use with or ordered by the armed forces of eight countries.

Ferranti Position and Azimuth Determining Equipment

UK

The Ferranti Position and Azimuth Determining System (PADS) has been designed by the Ferranti Inertial Systems Department to meet the British Army requirement for a compact, vehicle mounted, accurate survey and navigation system. The equipment displays continuously its position relative to a known starting point and provides orientation information relative to grid north. PADS, together with other equipment such as a laser rangefinder will assure first salvo effectiveness and fulfil the artillery survey requirement of providing gun locating data to within 10m (PE) for both Eastings and Northings,

The Position and Azimuth Determining System.

and azimuth to an accuracy better than 1mil. This accuracy is achieved for missions of duration of one hour and distances of 10km.

The sensing element is an inertia platform which measures accelerations acting on the vehicle in axes which are aligned and stabilised by high precision gyroscopes. The inertia platform feeds its outputs to a digital sensor computer which processes the signals and the resulting position and orientation data is displayed on a control and display panel of the PADS equipment.

Inertial navigators may be used in all types of vehicles and for high precision survey applications it is necessary to update the system periodically. The method of correction used is called 'velocity updating'. This entails stopping the vehicle at intervals of approximately 10min for a period of 10sec during which time the system errors are automatically eliminated. Position and orientation data are available as a continuous display and the ability to 'freeze' the display as required by the vehicle commander is provided.

Absolute orientation is in digital form enabling a bearing to be transferred optically to external equipment such as a gun director or laser range-finder. This orientation data is available continuously.

The main advantages of PADS, according to Ferranti, are that it is a completely self-contained system for installation in all vehicle types, has an accuracy to normal survey requirements, gives rapid automatic alignment and updating, and has built in self-testing facilities.

Employment

Production. In service with the British Army and other undisclosed countries.

Field Artillery Computer Equipment (FACE) UK

The FACE system is produced by Marconi Space and Defence Systems Limited and has been in service with the British Army since 1969. Its basic task is to automate the complex procedures associated with the preparation of gun-firing data for field artillery. Its three main advantages are: (a) Cut in training time. (b) Weapon response time reduced. (c) Greater accuracy. FACE can be fitted into a variety of vehicles including the FV432, FV610 and Land Rover. It can also be used in conjunction with AMETS or MILIMETS.

FACE can be used in two roles:

Artillery Role

In this role every piece of data that is required to place the shell on the target is displayed on the console, this data includes the gun and target position, allowances for the rotation of the earth, state of the weather, wear of gun barrels, variations in projectile weight and many other pieces of data. Full computation takes only a few seconds.

Survey Role

In this role more than 34 problems are stored each comprising a routine for input, the calculation, and an output routine for presenting the answer on the teleprinter. The calculations take less than 10 seconds. The system is based on the Elliott MCS 920B digital computer and uses a specially designed console for injecting the target and meteorological information, the console has four sections — data display, matrix, master controls and keyboard.

The complete system consists of the computer, console, power supply (and fan), supply control unit (and battery), inverter (power, static), line adapter, teleprinter and programme loading unit. It is silent and vibration free. The total weight of the system is some 300kg, power supply is 24V, 150W (stand by) and 300W (operational).

FACE installation in a Land Rover.

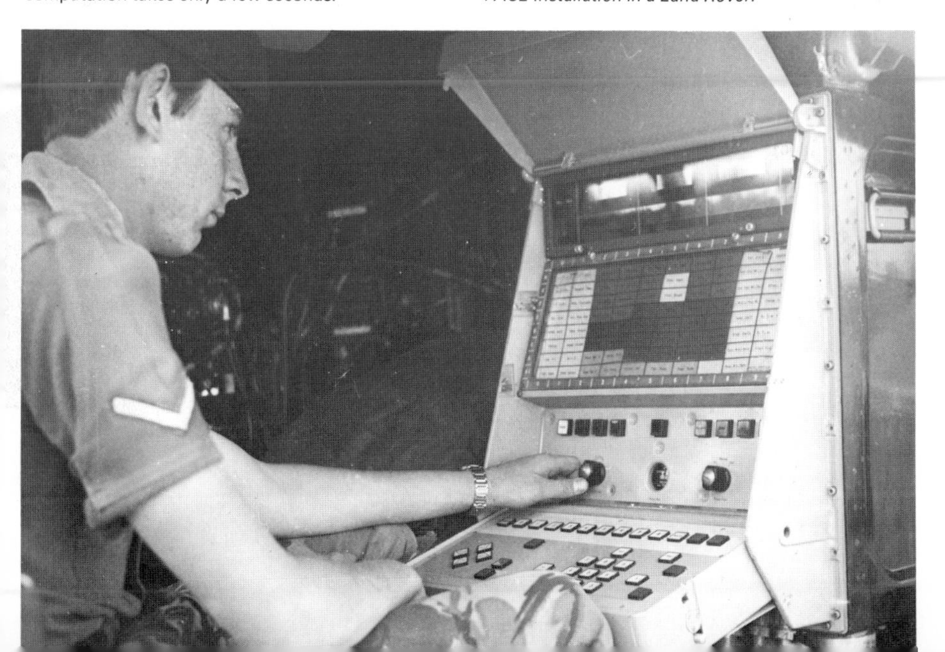

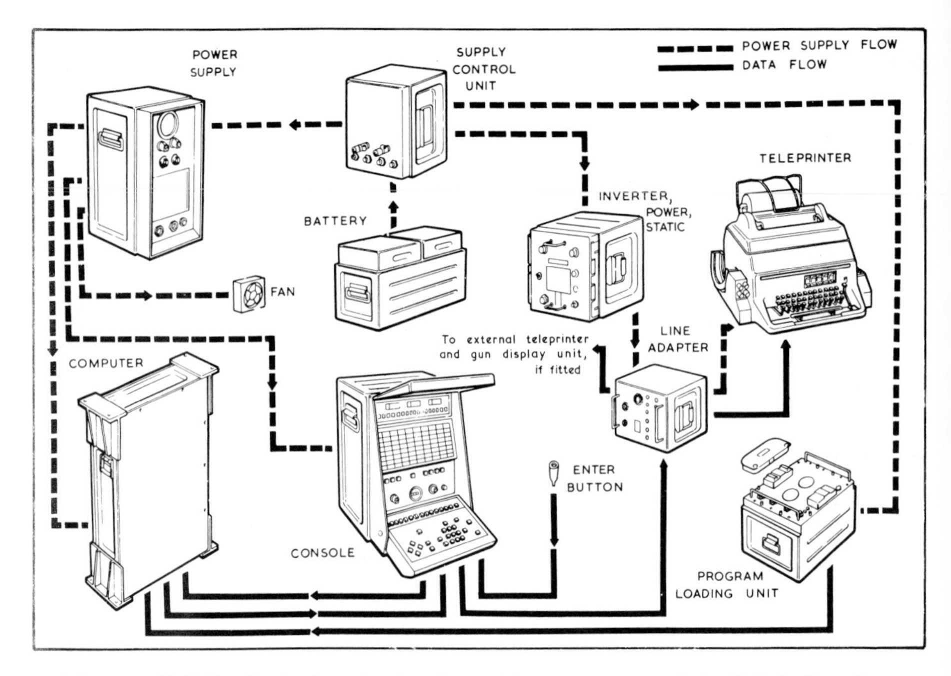

Main components of the Field Artillery Computer Equipment.

A Computer Field Test Set has been developed to enable unskilled operators to quickly locate faults. Once located the logic boards or sub-assemblies can be quickly replaced.

Battalion FACE
Battalion FACE uses the latest technology providing great storage, reliability and configuration options. It incorporates the MC 1800 computer which can produce firing data for up to 32 guns and transmits it to each gun's Visual Display Unit by line or radio. The MC 1800 computer is also a straight-forward plug-in replacement for the older 920B computer.

Employment
British Army, one FACE system is deployed with each battery. FACE is in service with more than 10 armies, these include Australia, Canada, Egypt and UK, and has been ordered by Chile.

BC 81 Artillery Computer — UK

This has been developed by Sperry Gyroscope to meet the requirements of the Swiss Army and uses the Sperry 1412 A computer which is already used in a number of other applications.

The BC 81 performs all the survey and ballistic computations that are essential to the effective fire control of field artillery.

At the artillery command post or fire direction centre there are only two compact units, the computer and the data input/output and display unit. Both units are designed for mounting in wheeled or tracked vehicles and can be quickly dismounted for use in buildings and field shelters. Alternatively the data input/output and display unit alone can be removed and operated remotely from the computer up to 30 metres away.

The computer unit embodies all the control and signalling facilities to transmit firing data automatically to each gun position via standard field cable or net radio. In towed artillery, the firing data is displayed for the gun detachment commander or on a ground-mounted unit, while in a self-propelled artillery weapon a master display and two slave displays are mounted in each self-propelled equipment.

AMETS Artillery Meteorological System — UK

This has been developed by Marconi Space and Defence Systems Limited and the Plessey Company Limited. Plessey developed the radio-sonde and the WF3M radar, the latter being developed from the civil WF3. AMETS is a self-contained mobile meteorological system designed for obtaining and processing information on upper air conditions.

It consists of a Command Post Vehicle (CPV), this tows the generator, $\frac{1}{4}$ton reconnaissance vehicle, this tows the radar trailer, optical tracker, stores vehicle, hydrogen trailers balloon filling shelter and meteorological balloons. These balloons consist of a radar reflector, radio-sonde, paper parachute and the balloon itself.

Basically a balloon is sent aloft, to this is attached the radar reflector and the radio-sonde transmitter, the balloon is automatically tracked by radar, while the radio-sonde transmits a signal conveying temperature. This signal is passed through a special converter which transforms the received radio-sonde signals into a form suitable for input into the computer. Data from the balloon tracking radar is also fed into the computer. Pressure is derived by the computer from radar height and surface pressure using standard ICAO atmosphere figures corrected for temperature. Average humidity values at each altitude are available on punched tape for all areas where AMETS operates. These values are fed into the computer store using the teleprinter. The computer used is the 920B, this is the same as that used in FACE.

The computer can provide eight types of meteorological messages ie standard artillery computer message (ie fed directly into FACE), standard ballistic message, surveillance drone message, sound ranging message, weapon locating radar message, biological/chemical message, nuclear fallout message and civil forecasting message. In addition 10 partial messages can also be produced during a sounding.

The CPV is the communications and control centre and contains the output teleprinter, the rear half of the vehicle also carries spare radio-sondes. The AIV contains the computer, radar display, teleprinter, digital display, sonde monitor, program loading unit, DPE console, computer test set, power supply test set and stand by batteries.

Both the AIV and the CPV are pallet mounted on Bedford 4ton 4 × 4 trucks and are air-conditioned and fitted with NBC equipment. AMETS can operate in temperatures ranging from — 32° to +52°C, and up to 20,000m.

Employment

Operational with the British Army.

AMETS deployed.

MILIMETS Artillery Meteorological System — UK

The MILIMETS meteorological system has been designed by Marconi Space and Defence Limited for the export market and allows highly accurate meteorological data to be computed and fed to field artillery computing equipment such as FACE or for use in manual computations. It is based on the successful AMETS equipment already proven in service with the British Army and in fact uses the same units but much more economically packaged. In place of the two 4ton vehicles required for AMETS the MILIMETS system uses two LWB Land Rovers to transport the complete system and to tow the radar. This allows high mobility, particularly across country and in rugged terrain.

The complete system consists of: (a) A data processing equipment vehicle. (b) A radar control equipment vehicle. (c) The Plessey WF3M tracking radar. The tracking radar is normally towed by the control vehicle with the data processing vehicle being used to tow a trailer carrying any ancillary equipment required, for example hydrogen cylinders for the balloons. A petrol generator, which is normally used to provide the power supply, is carried in the well of the radar vehicle and is sited outside when the system is deployed. The system will automatically revert to standby 24V batteries in the event of a generator failure without interruption to operation.

Each MILIMETS unit is a self-contained mobile

meteorological station using data processing equipment linked directly to a tracking radar. The well proven 920B computer is used in the system and is the same as that used in FACE and AMETS. The data processing equipment is virtually identical to that used in FACE which allows both equipments to have common spares, training, test equipment and base workshop facilities.

The main purpose of MILIMETS is to provide reliable meteorological information for firing batteries in a form which will enable quick response to a call for fire, with first salvo effectiveness. This information is obtained by tracking a balloon-borne radar reflector up to the altitude required. This data, together with temperature and pressure information is processed and passed to the various users in message formats suitable for the computation required. Messages are available for transmission from the MILIMETS system within seconds of the balloon reaching the required height.

MILIMETS deployed in the field.

The standard MILIMETS system does not use radio sondes, because in the warmer steady-state climates of the world, sufficient statistics are available to enable the computer to construct accurate temperature and pressure profiles from ground measurements. This method of interpolation is sufficiently accurate to ensure first salvo effectiveness in steady-state climates, and second salvo accuracy in less favourable circumstances. In regions where the weather conditions are not stable enough to justify this approach, the company can provide radio sondes and the associated ground equipment as additions to the standard MILIMETS system.

Employment

In service with undisclosed countries.

MORCOS Mortar Data Computing System — UK

MORCOS has been developed by Marconi Space and Defence Systems to calculate firing data for mortars. It is a self-contained unit incorporating the computer, keyboard for data entry, display and batteries. The case is made of durable plastic and profiled so that it fits comfortably into the hand. The unit is built to military specifications so that it is waterproof and will withstand all the environmental conditions likely to be encountered on the battlefield.

The full range of drills and computations required of the mortarman have been designed into a unit which is simple to use, inherently reliable, rapid and accurate in both initial fire prediction and subsequent corrections to bearing, elevation and fuze setting. MORCOS can produce the firing data for two mortar fire units simultaneously and rapid changes of mortar location can be achieved, eliminating the necessity for lengthy procedures previously required with the manual system. the MORCOS keyboard has only 24 keys, 10 are for the digits (0 to 9) and the remainder for a variety of functions.

The Marconi Space and Defence Systems' MORCOS computing system.

The MORCOS is simple to operate and after the operator has pressed one of the functional keys, for example mission, the integral computer leads him through the drills, calling for information to be entered either by displaying the parameters for which the operator must enter values or presenting on the display the names of the various drill routines for the operator to select whichever he wishes to use. At any time the operator can leave the drill in which he is engaged and return to the start point by pressing the EXIT key. Every entry made by the operator is displayed as he keys it in and can be cancelled by the CE (clear entry) key. After each entry has been checked and, if necessary, corrected, the operator presses the ENTER key, the computer accepts the entry and moves on to the next stage in the drill. Immediately all values with an abbreviated title are displayed in sequence, the computer moving on whenever the operator presses the NEXT key. MORCOS gives a range of simple drills, so that training is easy with a consequent saving in time and ammunition.

MORCOS operates on a 9V dc supply from standard commercial batteries, or from rechargeable batteries. No special first line test equipment is required. The complete system weighs only 0.91kg with batteries.

Employment

Production. In service with at least four countries.

Battlefield Artillery Target Engagement System (BATES) — UK

Late in January 1977 the Ministry of Defence Procurement Executive awarded a system definition contract for the Battlefield Artillery Target Engagement System (BATES) to Marconi Space and Defence Systems Limited. The study was carried out in three stages and was concerned with the use of computers to assist artillery staff in all areas of the battlefield. The first stage was carried out by teams from the company and the Ministry of Defence and identified those areas of artillery command and control which could be aided by automatic data processing. In stage two, appropriate parts of the system were examined to establish cost effectiveness and in the final, full definition of the system was carried out.

BATES will be a computerised command and control system which will combine with the target acquisition systems to enable artillery commanders to assimilate a very large amount of information and give them the power to utilise their field resources more efficiently and more effectively. A major feature of the system will be the use of distributed data processing which permits flexibility and power in artillery fire control systems. Each command level will have computing power appropriate to its function and call on information, via data links, from any other part of the system, mainly using radios designed and built by the company.

Employment

System definition.

Artillery Weapons Data Transmission System (AWDATS) — UK

This has been developed by Marconi Space and Defence Systems Limited. Each AWDATS consists of a digital coding unit situated at the command post, and a data display at each gun. Gun firing data from FACE is transmitted via the coding unit over existing radio or line equipment to the display units. This takes $2\frac{1}{2}$ seconds for a six gun battery. AWDATS is in service with the British Army.

AWDATS artillery weapon data transmission system in use.

Cymbeline Mortar Locating Radar

UK

Cymbeline (Radar Field Artillery No 15 Mk 1) has been developed by EMI Electronics Limited, in association with government establishments since the 1960s as a replacement for Green Archer.

Cymbeline is designed not only for mortar location duties but also for coastal surveillance, helicopter and light aircraft control, rapid survey, meteorological balloon tracing and the adjustment of fire by air bursts and ground bursts.

Its main features are its light weight, high accuracy of location and adjustment, high tactical mobility, high reliability, easy to operate and maintain, built in self-test facilities, built in target simulator for training purposes and it is very quick into action.

Cymbeline is similar in operation to Green Archer, it does however have a third radar beam elevation angle that may be used to alert the operator. The radar unit itself consists of three basic groups of assemblies — (a) The mounting, (b) the radar head and, (c) the remote control and indicator units, these can be operated up to 15m away from the radar. Unlike the Green Archer, Cymbeline has its own built-in generator set, this is inaudible beyond 200m and it is powered by a Wankel engine. It is provided with 4 manual screw jacks for levelling purposes.

Its maximum displayed range is 20km, minimum detection range is 1km, maximum detection range depends on the calibre of the mortar, location time is 15-20sec. Total weight of the radar and the trailer is 980kg. Normal operating crew is two men, although it can be operated by one man.

The display unit contains a built-in radar simulator which provides realistic synthetic signals to the cathode ray tube representing intercepts of mortar bombs. All main units contain built in test circuits by means of which the radar operator can locate a defective unit, this can then be quickly replaced. A data memory storage unit is available as an optional extra, this allows video signals from the receiver to be stored and displayed on the cathode ray tube.

Cymbeline can be transported in four ways. First by mounting on a two-wheeled one tonne trailer towed by a long wheel base Land Rover, second carried by a helicopter (ie a Wessex or Puma) third it can be quickly detached from the trailer and carried by four men using lifting poles and slings, fourth mounted on the roof of the FV432 APC (or similar vehicle). In the latter case it is fitted with an automatic hydraulic levelling system. The Mk 1 is towed whilst the Mk 2 is mounted on the FV432 APC.

Cymbeline in operational position complete with remote control and indicator units.

Employment

In service with the British Army from 1973 and other undisclosed armies.

Green Archer Mortar Locating Radar

UK

The Green Archer system was developed by EMI Limited in association with government establishments from the mid-1950s. The first prototype was built in 1958 and the system entered service with the British Army in 1962. The role of Green Archer is the location of enemy mortar positions but it can also be used for the adjustment of artillery fire, surveillance duties and the control of light aircraft and helicopters.

The basic idea is that the radar operator plots two points on the mortar bomb trajectory after it leaves the weapon. The slant range and bearing to each of the plotted positions is measured, in addition the flight time between the two points is measured. The position of the mortar is then determined by the analogue computer from this information plus the pre-set elevation angles of the radar. It takes some 30sec to determine the position of the enemy mortar. Max range is reported to be 17,000m.

The Green Archer has been replaced in the British Army by the much lighter Cymbeline mortar locating radar system, also developed by EMI. The British Army used two models, trailer mounted (Radar Field Artillery No 8 Mk 1) and mounted on a special version of the FV432 called the FV436 (Radar Field Artillery No 8 Mk 11), The German and Danish armies have used the system mounted on M113 APCs.

The Green Archer radar trailer.

Radar Field Artillery No 8 Mk 1

This is the trailer mounted version. The radar is mounted on a four-wheeled trailer and there is a separate trailer for the silenced generator. Basic data of the radar trailer is: weight 2,915kg, length of trailer 4m, length of trailer including tow bar 6m, width 2m, height travelling 2m, height operational 2.9m and ground clearance is .38m. Both the radar and generator trailers are towed by a 1ton armoured truck, or similar vehicle.

The following training aids have also been built:
Simulator, Radar, Jamming, No 2 Mk 1
Simulator, Radar, Target, No 3 Mk 1
Simulator, Radar, Target, No 4 Mk 1

There are 13 major sub-assemblies in Green Archer, and the fault can be quickly detected and replaced by the spares vehicle, this vehicle carries a complete set of spare parts and sub-assemblies. There is one of these vehicles in each Green Archer detachment.

Employment

Sales of the Green Archer were made to Denmark, Italy, Israel, South Africa, Sweden, Switzerland, and the UK, some of these have replaced it with the Cymbeline.

Tellurometer MR5A Distance Measuring System UK

The MR5A microwave electronic distance measuring instrument has been specially designed and produced by Tellurometer, a Plessey company, for military applications. The MR5A indicates a measured distance directly in metres and centimetres, operation being either fully automatic or manual at the choice of the operator. It is capable of indicating unambiguous display of ranges up to 80km or down to 100mm with a probable error, using a series of 'fine' readings, of less than 10mm. Scale error due to refractive index determination is usually less than five parts per million and actual measurement can be accomplished in less than 20sec.

A recent innovation is the introduction of facilities for separating the antenna unit from the control unit by as much as 25m; the antenna can then be mounted on a specially developed mast, thus permitting direct ranging between points which might otherwise be screened from each other. A vehicle installation kit is available to allow the system to be fitted into a Land Rover FFR.

Another Tellurometer system is the MRB201 which is a dynamic distance measuring system for applications involving ranges in excess of 50km and up to 200km in the airborne role. This has already been adopted by the French Army.

The Tellurometer MR5A distance measuring equipment in service with a British Royal Artillery unit armed with 105mm light guns.

Employment

The MR5A is used by many countries including Canada, Thailand and the UK.

Plessey No 2 Mk 1 Sound Ranging Link UK

The Sound Ranging Radio Link No 2 Mk 1 has been developed by Plessey Avionics and Communications to meet the requirements of the British Royal Artillery. It comprises a base of up to seven microphone positions connected by a radio data link to a data-recording and processing centre. The system records the time differences of gun sounds reaching adjacent microphones, from which bearing information is produced to provide the target location. To achieve accurate locations the position of each of the microphones must be fixed by precise survey, and the time intervals must be corrected for non-standard atmospheric conditions.

An observer is deployed in the forward area; it is his task to listen to the enemy artillery fire and to switch the system on when he wishes to record. He also provides additional information such as estimated type and calibre of weapon, number of rounds fired etc. This information is relayed to the data-recording and processing centre and the position of the hostile weapons established. In the British Army the system is deployed in a modified FV432 APC.

Employment

Great Britain and other undisclosed countries.

Plessey sound ranging link No 2 Mk 1 mounted in a FV432 APC.

DR810 Muzzle Velocity Radar USA

The DR 810 muzzle velocity radar was originally developed by the Danish company of B&W Elektronik as a successor to their earlier DR513 muzzle velocity radar. Subsequently the Danish company licenced the American company of Lear Siegler Incorporated (Astronics Division) to manufacture the equipment in the United States. The DR810 is marketed on a worldwide basis by the Avitron Division of Lear Siegler Incorporated.

The DR810 muzzle velocity radar has been designed for installation on field and anti-aircraft artillery and also has naval applications. The system consists of an antenna unit, antenna mounting bracket, data unit, interconnecting cable and drum assembly, power cable, test unit, accessory bag and a transport box.

The transmitting antenna transmits a high frequency signal along the line of sight of the muzzle bore. A small part of the signal is reflected from the projectile and received by the receiving antenna. The frequency of the transmitted signal is compared to the frequency of the received signal in the receiver unit. The difference between the transmitted and received signal is a low frequency doppler signal, the frequency of which is proportional to the velocity of the projectile. This low frequency doppler signal is fed to the data unit via the interconnecting cable.

The velocity range for the measurements system in the data unit is divided into 16 selectable ranges covering velocities from 50 to 1,750 m/sec. Each of the 16 ranges has a bandwidth of 100m/sec which provides a 100m/sec overlap between ranges. The muzzle velocity is given directly in metres a second on the front panel display.

The DR810 is suitable for weapons ranging in calibre from 30 to 205mm and is accurate to 0.1%. The antenna unit weighs 9kg while the data unit weighs 23kg, optional equipment includes a tripod when it is not possible to mount the antenna directly on the weapon, and a mains unit.

Employment

In service with the United States Army.

Main components of the DR810 muzzle velocity radar.

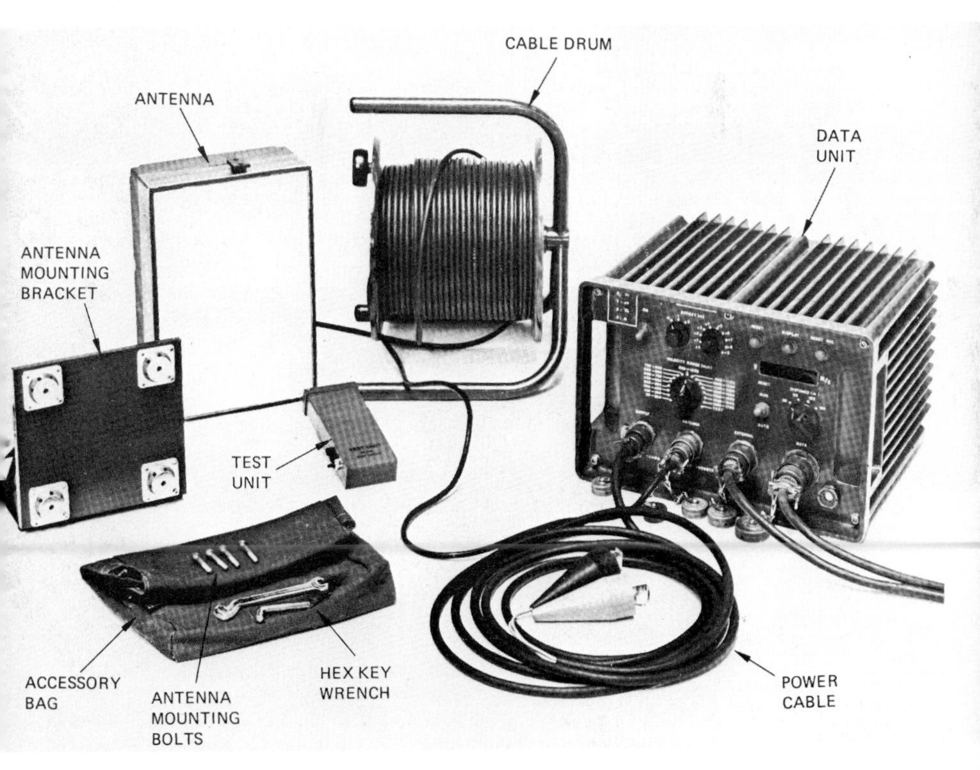

RCA AN/GVS-5 Laser Rangefinder — USA

The AN/GVS-5 handheld laser rangefinder was developed by the RCA Government Communications and Automated Systems Division under contract from the United States Army Electronics Command, Fort Monmouth, New Jersey. Twenty prototypes were built by the company under a $1.5million contract awarded in April 1975. A production contract worth $10.7million for 1,500 rangefinders was announced in 1978.

The AN/GVS-5 handheld laser rangefinder has been developed to enable forward observers and mortar teams to obtain fast and accurate measurement of distance to the target. It consists of three main groups, transmitter, optics and receiver plus the power source.

The transmitter group includes the laser transmitter module, pulse-forming network (PFN), capacitor and inductor, and the trigger circuit module. Contributing to the low cost of the rangefinder is a chemical-dye Q-switch wafer, which turns transparent at the correct power level to produce the laser pulse. This switch combines with the reflecting mirror and the laser rod to form the resonator sub-assembly. The resonator sub-assembly, flashlamp, trigger wire and structure comprise the laser transmitter module, which is pre-aligned at the factory for simple replacement in the field.

The optical group combines the transmitter and receiver optics in one integral casting. The three major sub-assemblies are the housing, transmitter telescope and the receiver/sighting telescope. The latter is essentially like half a binocular assembly with an objective lens and an eyepiece. A beam splitter in the optical path provides a means for projection of a reticle and LED range display to the eyepiece, and for providing an optical path to the photodiode detector for the returned laser energy.

The receiver group includes the detector/preamplifier module, video amplifier module, and range counter/display module. A silicon avalanche diode is used as the detector; it and the preamplifier are included in one hybrid circuit sealed in a small package. The associated video amplifier is also a hybrid providing electronically controlled video gain. Two silicon-on-sapphire CMOS circuits, a crystal oscillator and a display sub-assembly comprise the essential part of the range counter/display module hybrid. A range gate, continuously adjustable to

RCA AN/GVS-5 handheld laser rangefinder.

5,000m can be employed to gate out close-in indesired targets. Range to the target is displayed numerically by light-emitting diode indicators. The latter also provide indication of low battery voltage and the presence of multiple targets.

Power is provided by a rechargeable NiCd battery or an external power source, with the former providing sufficient power for at least 700 shots.

Employment

United States

AN/PAQ-1 Laser Target Designator

USA

The AN/PAQ-1 Laser Target Designator (or LTD for short) has been developed for the United States Army Missile Research and Development Command by the Hughes Aircraft Corporation. Following successful trials with prototype LTDs a production order was placed for 152 units for the Army and a further 25 units for the Air Force, total value of the contract was $15 million. First production units were delivered in 1978 and the second year contract calls for the production of a further 112 units.

The LTD can designate targets for any of the US Forces' tri-service laser homing weapons, mark the position of troops or designate sides for aerial supply drops.

The LTD directs an invisible beam of laser pulses at the target, these are reflected from the target and can easily be detected by special sensors in laser homing missiles or projectiles.

The equipment consists of three easily-replaceable modules designed to withstand field handling. The battery pack, which makes up the stock of the LTD and contains 22 battery cells which can be quickly replaced as one pack. The second component, the power supply, transforms the battery

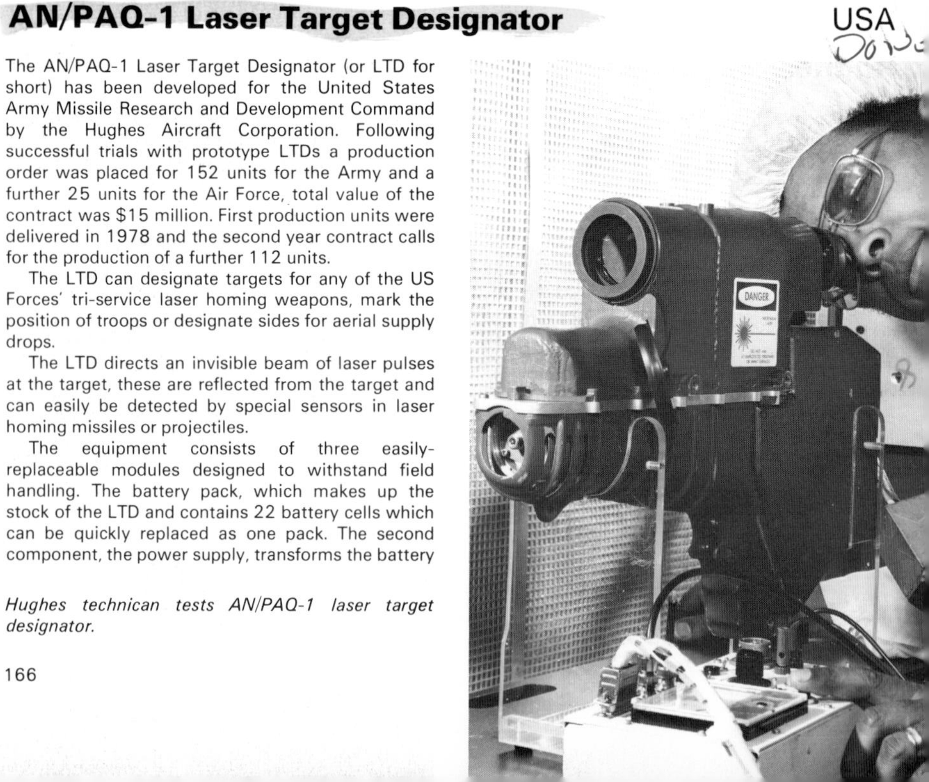

Hughes technican tests AN/PAQ-1 laser target designator.

current into energy useable by the laser transmitter, the third component.

Employment

In service with the United States Army and Air Force.

AN/TVQ-2 Ground Laser Location Designator USA

Done

The AN/TVQ-2 Ground Laser Locator Designator (or GLLD for short) has been developed for the United States Army Missile Research and Development Command by the Hughes Aircraft Company. The GLLD is a lightweight portable tripod-mounted system which enables a ground soldier to precisely locate and designate targets for attack by armed helicopters, laser-homing weapons (such as the Copperhead CLGP, laser Maverick and the Hellfire) or conventional artillery fire.

The system consists of a laser designator, laser rangefinder, self-contained batteries, a tripod and a viscous damped tracking unit. It weighs a total of 22.7kg and can easily be moved to a forward position by two men. The viscous damped tracking unit provides the accuracy to work against rapidly moving distant targets.

When a target is sighted, the operator uses the GLLD's laser rangefinder to determine its distance and bearing, this information can then be relayed to conventional artillery for effective shelling or to assist aircraft or helicopters for close support. This information may also be provided to remote artillery or aircraft provided with laser-homing projectiles or missiles. In this case, the GLLD operator focuses a narrow-beam, high-intensity laser on the portion of the target he wants the laser-homing munition to strike. While in flight these weapons sense the laser reflected upward off the target and guide themselves down this cone of invisible pulsed light to a hit. The wavelength and periodic pulsation of the laser beam allows the weapon to differentiate the correct target from others designated by different GLLD units in the same battlefield area.

Hughes Ground Laser Locator Designator in a battlefield environment.

Following extensive field tests and evaluation with 22 prototype GLLDs, the US Army placed a production order with Hughes for 130 units.

Hughes Modular Universal Laser Equipment USA

Under a contract worth $6million, the Hughes Aircraft Company has built six engineering development models of the Modular Universal Laser Equipment (MULE for short). These were delivered for six months of operational tests in the summer of 1979. The contract is being managed for the United States Marine Corps by the United States Army Missile Research and Development Command at Redstone Arsenal.

MULE consists of three modules, the laser designator/rangefinder module, a north finding module, and a stabilised target tracker module which is a multi-functional tripod. The designator/rangefinder, which resembles a short barrelled rifle, can be detached from the tripod and hand aimed for target designation or rangefinding. The tripod displays range, azimuth and angle of elevation of targets, and provides a viscous liquid damping platform for precisely tracking moving targets. The

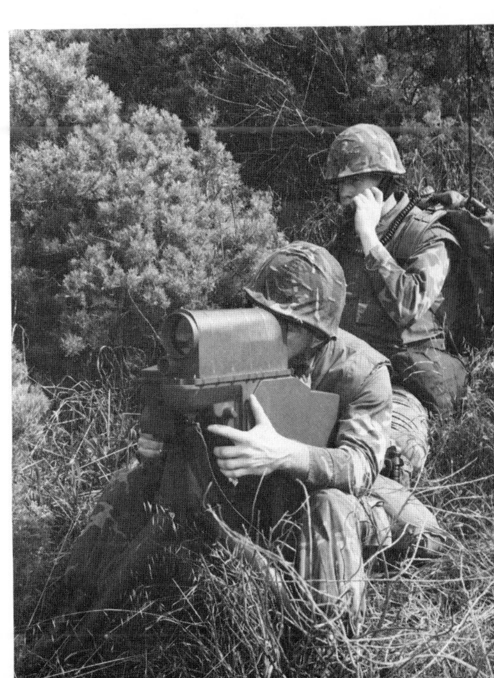

The Hughes Modular Universal Laser Equipment undergoing trials.

north-finding module which is a small gyro compass which locates true north to provide azimuth accuracy, has been developed under the management of the Naval Weapons Centre at China Lake.

MULE will be used to designate targets for all laser guided weapons now operational or under development including CLGPs, laser Maverick or Hellfire. It also has the capability of combining azimuth, elevation and range information into a digital message that can be sent through a Digital Communications Terminal to an automatic tactical fire control radar.

The system has parts in common with the AN/PAQ-1 Laser Target Designator and the AN/TVQ-2 Ground Laser Locator Designator tracking tripod, both built by Hughes, and the AN/GVS-5 hand held laser rangefinder developed by RCA. MULE components such as the eyepiece, resonator optics, test equipment and electronic modules will be common to and interchangeable with these systems.

Norden AN/GYK-29 Battery Computer System — USA

In 1975 Marconi Space and Defence Systems of the United Kingdom signed an agreement with the Norden Division of United Technologies under which Norden could propose to the United States Army an artillery battery computer system using Marconi-developed equipment. This proposal was accepted by the United States Army and prototypes are now undergoing field evaluation. The battery computer system will replace the obsolete Field Artillery Digital Automatic Computer (FADAC) and also extend the capability of TACFIRE.

The battery computer system integrates key battlefield artillery echelons for rapid battlefield command and control. It provides direct digital access to forward observers and to battalion TACFIRE and also operates in an autonomous mode. It accepts fire requests from forward observers, automatically computes firing data, and displays fire commands at each weapon. As many as 12 guns or howitzers can receive their individual ballistic computations simultaneously from the BCS. Up to three separate, concurrent fire missions can be executed with the system. It is also capable of computing fire commands for other weapons such as the Multiple Launch Rocket System and Lance.

The heart of the Norden BCS is the battery computer unit. This maintains a digital communication link with fire direction officers, fire support officers, forward observers, and weapon section chiefs. It performs all required computations for first-round accuracy, with the results appearing on a 1728 character plasma display. The central processor within the BCU contains 32K × 24 Bit words of memory, with error detection and correction circuitry for fail safe reliability. The design incorporates growth capacity of 50% in memory.

THE BCU houses a magnetic tape cartridge for software program and data storage and this interfaces with existing American COMSEC equipment and printers, data is transmitted to the guns in one second.

The Gun Display Unit is the final link in the BCS communications chain. Its place is with each weapon in the firing battery. The GDU consists of three assemblies, a Section Chief Assembly (SCA) and two Gun Assemblies (GA), plus a carrying case that also serves as a signal and power distribution unit. The SCA is a personal tool for the section chief, the man who commands the gun crew. It is hand held and connected by cable to the carrying case thus giving instant access to all gun related data and commands. The device contains its own memory bank for sequential or direct access to stored fire mission information. It is equipped with audible alarms and visual cueing to alert the section chief to fire missions and check fires. A bright eight-digit display shows an easily read, positive indication of gun commands under all conditions and a single button sequences the display of gun orders.

Norden Battery Computing system.

The two identical gun assemblies on the weapon show the gunner and assistant gunner separate elevation and deflection displays. They are connected by wire to the carrying case (with growth capability for radio) and receive the commands directly.

Employment

Early in 1980 Norden was awarded a contract for 687 BCS to be delivered over a five-year period.

TACFIRE Artillery Fire Direction System — USA

TACFIRE is a total command and control system for the field artillery and not only encompasses the battalion FDC and its sub-elements, but adds a division FDC and a Fire Support Co-ordination Element as required for larger Army operations. The requirement for TACFIRE was written in 1966 and in 1967 the contract was awarded to Litton Industries (Data Systems Division).

All the equipment is housed in man-transportable transit cases and can be mounted in S-280 shelters for deployment. TACFIRE improves the accuracy, capability and time responses of the field artillery.

TACFIRE consists of a central Fire Direction Centre (FDC) at Division Artillery and at Battalions with a remote device at batteries, the Battery Display Unit (BDU) and the Variable Format Message Entry Device (VFMED) with the Fire Support Officers (FSO), Fire Support Elements (FSE), Forward Observers (FO) and Operations Centres (OC).

The battalion level TACFIRE consists of an AN/GYK-12 computer, power converter group, data terminal unit, module test unit, communications control unit, removable media memory unit, artillery control console, digital plotter map, electronic line printer and the remote control monitor unit, all of which is mounted in one shelter.

At the division artillery level the system consists of the same units as the battalion level plus one additional mass core memory unit, a dual removable media memory unit, an electronic tactical display and an auxiliary data terminal unit, all of which is mounted in two S-280 shelters.

The two remote equipments are the variable format message entry device and the battery display unit. The former consists of a remote data terminal, display editor, keyboard and an electronic line printer. The battery display unit consists of an Electronic Line Printer and a remote data terminal.

There is also the Electronic Tactical Display (ETD) at the division FDC to analyse intelligence data. Another feature of TACFIRE is that the survey and Met information can also be fed into the system. TACFIRE can also perform the job of Preliminary Target Analysis, ie the capability of artillery to defeat the target, or should aircraft or armour be used.

Electronic tactical display.

Artillery control console.

Employment

The US Army accepted its first TACFIRE system for full scale development in October 1980 and the first non-test systems should be delivered in 1980. Earlier in 1978 the General Accounting Office said that the digital transmitters used by the forward observers were successful only 28% of the time.

AN/TPQ-36 Mortar Locating Radar — USA

The AN/TPQ-36 mortar locating radar has been developed to meet the requirements of the US Army and Marine Corps by the Ground Systems Group of the Hughes Aircraft Company, under the direction of the MALOR Project Management Office of the United States Army Electronics Command. An Army division has a total of three AN/TPQ-36 mortar locating radars and two AN/TPQ-37 artillery locating radars, with the complete system being called the Firefinder. In 1978 a contract was awarded to Hughes for 106 AN/TPQ-36s at a total cost of £166million.

The AN/TPQ-36 consists of two major components. First the antenna-transceiver trailer which carries a turbine power generator, the radar transmitter and receiver and second the operations control shelter which contains the balance of the processing equipment, the weapons location unit and communications equipment. The system is towed by an M561 'Gama Goat' and can be deployed in 15min and moved in 5min.

In action the unmanned antenna trailer is sited behind masking terrain in a position providing camouflage and good radar coverage of the air above the hostile horizon. The manned operations control shelter can be positioned up to 100m away from the antenna. The antenna provides coverage through a 90° azimuth sector from each stationary position. From the operations control shelter, the operator can tilt the power-driven antenna to provide horizon clearance, or rotate it to any azimuth position. In addition, the system offers a 270° sectoring mode in which it will automatically search one sector for a short period, track and locate targets in it and then turn to the next sector.

Although computer software and some hardware on the two radars differ, the principle of operation is the same. A 'fence' of pencil-shaped radar beams, adjustable according to terrain, is swept along a 90° section of the horizon several times each second. As any object breaks upward through the fence, the system instantly and automatically transmits a verification beam. If it, too, detects the target, the computer initiates a rapid succession of tracking beams at a much higher data rate. While tracking this target, the radar continues to scan, locate other targets and develop tracks on them.

As the beams track a target, the signal and data processors test each track against a series of discriminants to filter out unwanted targets. When the computer establishes that a target is valid, it smooths the measured track data and derives a trajectory, which it extrapolates to establish the firing position. This is displayed on a visual map and printed out in map coordinates, which may be interfaced automatically with a tactical fire control system such as the United States Army Tacfire.

AN/TPQ-36 with antenna unit on left and operations control shelter mounted in the rear of the M561 'Gama Goat' (6×6) vehicle.

The system's automatic detection, tracking and location process is so rapid that the position of the weapon which fires a round can often be determined before the round itself reaches the target. System accuracy is consistent with the accuracy and effective radius of counterfire weapons.

In addition, both radars can also track the fire of friendly weapons, predict its point of impact and provide registration and adjustment. Both can determine priorities which enemy guns are hitting which targets, thus providing information to determine for counterfire.

Employment

In service with US Army and Marine Corps.

AN/TPQ-37 Artillery Locating Radar USA

The AN/TPQ-37 Artillery Locating Radar has been developed to meet the requirements of the United States Army and Marine Corps by the Ground Systems Group of the Hughes Aircraft Company. An Army Division has a total of three AN/TPQ-36 mortar locating radars and two AN/TPQ-37 artillery locating radars, with the complete system being called the Firefinder for short. The AN/TPQ-37 entered service with the US Army in 1978. The Army has a total requirement for 70 systems of which 32 have so far been ordered, one batch of 10 and one of 22.

The AN/TPQ-37 has a range in excess of 30km and employs the same operations control shelter as the smaller AN/TPQ-36, but the radar radiates more power through a larger antenna to cover a larger area. The antenna-equipment trailer is fitted with detachable wheels and is normally towed by a 5ton 6 × 6 truck which also carries the 400Hz generator which supplies electrical power to the system. The complete system can be emplaced in 30min and be moved in 15min. The system of operation is almost identical to that of the AN/TPQ-36 and is fully described in the latter entry.

AN/TPQ-37 artillery locating radar during trials at Fort Sill.

AN/MPQ-4A Mortar Locating Radar

USA

Prototypes of the AN/MPQ-4A were designed and built by the United States Army Electronics Command at Fort Monmouth, New Jersey. It was designed to replace the AN/MPQ-10 which was used during the Korean war. Production AN/MPQ-4As were built by the Heavy Military Electronics Division of the General Electric Corporation of Syracuse, New York.

The radar is mounted on a two wheeled trailer and towed by a standard 6 × 6 2.5ton truck. Power is supplied by a three-phase, Y-connected gasoline generator which is also mounted on a separate two-wheel trailer. The radar set can be operated by remote control and in addition to its role of locating enemy mortars it can also adjust friendly artillery fire to a maximum range of 10,000m.

The Foster dual scanner and built in computer allow the radar to provide a readout of the coordinates of enemy mortars or artillery when only one projectile has passed through the radar's dual beams. The sector scan of the blade type antenna is designed to search electronically in azimuth and mechanically in elevation.

AN/MPQ-4 mortar locating radar of the Japanese Ground Self Defence Force (their designation of the system being the JAN/MPQ-N1).

AN/MPQ-49 Forward Area Alerting Radar

USA

The Forward Area Alerting Radar was developed under contract to the United States Army Missile Command by Saunders Associates. Initial production, from 1971/72 was by Saunders Associates but in 1974 Sperry Gyroscope were awarded the second production contract. The FAAR is used to provide early warning to the 20mm Vulcan anti-aircraft gun systems (towed and self-propelled), as well as the Chaparall, Redeye and Stinger SAMs. Two versions are currently in service, the AN/MPQ-49 (mounted in the rear of an M561 Gama Goat vehicle) and the AN/MPQ-54 (mounted on an M796 trailer) which is towed by a standard M35 truck or similar vehicle.

The FAAR is an L-band pulse doppler radar utilising a low noise stable master oscillator, grid-modulated transmitter and receiver. The latter translates target information to video frequencies which are range gated, doppler filtered and compared to an alarm threshold. Target data is then transmitted to the Target Alert Data Display Set (TADDS) via the RF data link.

The TADDS, which is designated the AN/GSQ-137, is a self-contained battery operated device deployed with the weapons. Aircraft are indicated on the display matrix by either of two discs in each matrix square, green for friend and red for foe aircraft moves into an adjoining square, the weapons operator maintains a track of the target movement on the face of the unit. The gunner then acquires the target and when this is in range opens fire.

Employment

US and other armed forces.

AN/MPQ-49 forward area altering radar on M561 'Gama Goat' (6×6) vehicle.

Photo Credits

The photographs used to illustrate *Artillery of the World* have come from many companies, governments and individuals all over the world. The sources, where known, are listed below.

AEG-Telefunken, Germany
APN
Argentine MoD
Austrian MoD
B and W Electronik, Denmark
Bennett, R. M.
Bernardini, Brazil
Bofors, Sweden
CILAS, France
Contraves Co, Italy and Switzerland
Ebata, K.
Electronique Marcel Dassault, France
Esperanza Y, Cia, SA, Spain
Ferranti Ltd, UK
Finnish MoD
Gander, T. J.
General Electric Co, USA
German MoD
GIAT France
Hollandse Signaalapparaten, Netherlands
Hughes Co, USA
Indian MoD
Israel Defence Forces (General Staff)
Israel Aircraft Industries
Leaford, M.
Kongsberg Vapenfabrikk, Norway
Litton Systems Inc (Data Systems Division), USA
L. M. Ericsson Co, Sweden
Marconi Space and Defence Systems Co Ltd, UK
MoD (Army), UK
MOWAG, Switzerland
Nera Bergen, Norway
Nogi, K.
N.V. Optische Industrie, Netherlands
Oerlikon Bührle Co, Switzerland
Philips Teleindustri AB, Sweden
Plessey Co
Radatronic AS, Denmark
Rheinmetall Co, Germany
RAFAEL Armament Development Authority, Israel
Saab-Scania (Aerospace Division), Sweden
SEP, France
Saunders Associates, USA
Simrad AS, Norway
Soltam Co, Israel
Steyr Co, Austria
Swedish Army Material Department
Swiss MoD
Taibo J.I., Spain
Tass
Thomson-Brandt (Armament Department), France
US Army
Vought, USA

Index

Equipment

Guns

Mortars

Multiple Rocket Systems